Innovation in Pedagogy and Technology Symposium, 2019

Advancing Technology in Education at the University of Nebraska

May 7, 2019

Selected Conference Proceedings

Presented by University of Nebraska Online and
University of Nebraska Information Technology Services

Zea Books
Lincoln, Nebraska
2020

Nebraska UNIVERSITY OF
Lincoln®

Nebraska. UNIVERSITY OF
Online

Nebraska UNIVERSITY OF
INFORMATION TECHNOLOGY SERVICES

ISBN 978-1-60962-168-1

doi 10.32873/unl.dc.zea.1099

Zea Books are published by University of Nebraska-Lincoln Libraries.

Composed in Segoe types.
Zea Books are published by the
University of Nebraska–Lincoln Libraries
Electronic (pdf) edition available online at
https://digitalcommons.unl.edu/zeabook/
Print edition available from
http://www.lulu.com/spotlight/unllib
UNL does not discriminate based upon any protected status.
Please go to http://www.unl.edu/equity/notice-nondiscrimination
Symposium website:
https://symposium.nebraska.edu/agenda

Contents

Advancing Technology in Education
at the University of Nebraska

Technology has forever changed the landscape of higher education, and continues to do so—often at a rapid pace. At the University of Nebraska, we strive to embrace technology to enhance both teaching and learning, to provide key support systems and meet institutional goals.

The Innovation in Pedagogy and Technology Symposium is for any NU administrator, faculty and staff member who is involved with the use of technology in education at all levels. The 2019 Symposium drew more than 500 NU faculty, staff, and IT professionals from across the four campuses for a day of discovery and networking.

This year's Symposium was held at the beautiful Cornhusker Hotel in downtown Lincoln. Dynamic conversations on topics in education and technology included presentations by University of Nebraska faculty, staff and administrators, concurrent sessions focused on emerging technologies, pedagogy/instructional design and support and administrative strategies, panel discussions, roundtable discussions and networking time, and sponsor exhibits. Continental breakfast and lunch were provided.

2019 Keynote and Featured Speakers

Bryan Alexander, Ph.D.
Futurist, Researcher, Writer, Consultant & Teacher

Tanya Joosten, Ph.D.
Educator, Author, Thought Leader & Social Scientist

Victoria Brown, Ph.D.
Researcher, Leader, Educator & Critical Thinker

Program

Welcome Address

Susan Fritz, Ph.D.,
Executive Vice President and Provost, University of Nebraska

Opening Remarks

Mary Niemiec
Associate Vice President for Digital Education,
Director of University of Nebraska Online

Keynote Presentation

Shaping the Next Generation of Higher Education

Bryan Alexander, Ph.D.

What will the university become over the next generation? During the keynote address we explore the leading forces reshaping higher education. We begin with macro social forces, including demographics, economics and culture. Next, we will survey technological developments, starting with the established ones (mobile, social, gaming) and move on to emerging trends (mixed reality, artificial intelligence). With the context set, we move into academia itself, looking into enrollment, financing and strategy changes. In conclusion, we connect multiple forces, exploring emergent educational technologies.

Featured Extended Presentation

Redesigning Courses & Determining Effectiveness Through Research

Tanya Joosten, University of Wisconsin-Milwaukee (UWM)
Erin Blankenship, Ph.D. (UNL)
Ella Burnham (UNL)
Nate Eidem, Ph.D. (UNK)
Marnie Imhoff (UNMC)
Linsey Donner (UNMC)
Ellie Miller (UNMC)

The dialectical between practice and research is paramount. The design of online courses should be informed by research and online courses should be researched and constructively evaluated to inform future instruction of courses to improve the quality of online learning. Teams of faculty and instructional support staff across Nebraska redesigned their courses for the online environment. Additionally, they designed research in order to determine if their course was effective and to identify which components of their course worked and which needed to be improved. Others designed research to better understand their students' success in the online environment. This panel of participants will discuss their experiences and lessons learned about the process and their courses, share tips and resources on redesigning courses and conducting research and highlight obstacles that they had to overcome and advice for others. Stick around for the second half of the presentation when panelists collaborate with audience members to discuss upcoming course redesign projects.

This presentation featured:
- the importance of research in designing online courses.
- real-world examples where this has improved online courses.
- the process in redesigning these courses.

---

5 Ways to Utilize Canvas Data

Ji Guo, Ph.D. (UNL), Jessica Steffen (UNCA)

The implementation of learning management systems such as Canvas have provided universities with a critical resource for gathering and analyzing data related to these questions. These evolving data capabilities better allow for the examination of current as well as emerging topics within the fields of teaching and learning, instructional design and academic assessment amongst others. This session will consequently discuss the purpose of learning analytics and its current state at UNL through the use of five Tableau dashboards on such relevant topics as student engagement, learning outcomes, at-risk student identification and overall student performance. By utilizing these dashboards, faculty will ideally be able to utilize Canvas data for decision-making pertaining to their curriculum, course design and/or teaching methods.

This presentation featured:
- learning analytics initiatives and capabilities developing at UNL.
- course-level data models pertaining to concepts of student engagement, at risk-student factors and overall student achievement.
- connections between various forms of course-level data and resulting academic behavior and/or performance.

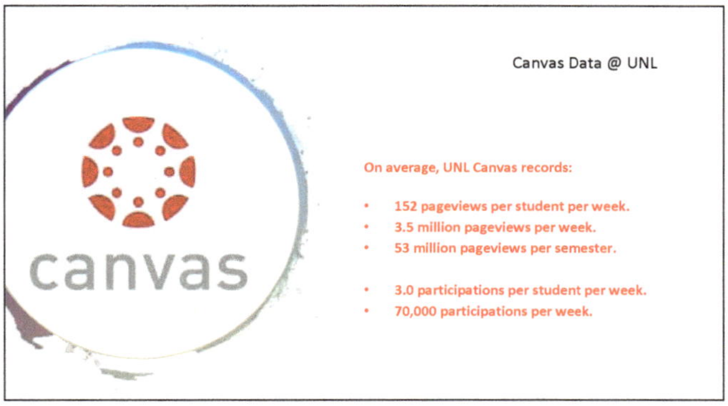

Slide 1

Canvas Data @ UNL

On average, UNL Canvas records:

- 152 pageviews per student per week.
- 3.5 million pageviews per week.
- 53 million pageviews per semester.

- 3.0 participations per student per week.
- 70,000 participations per week.

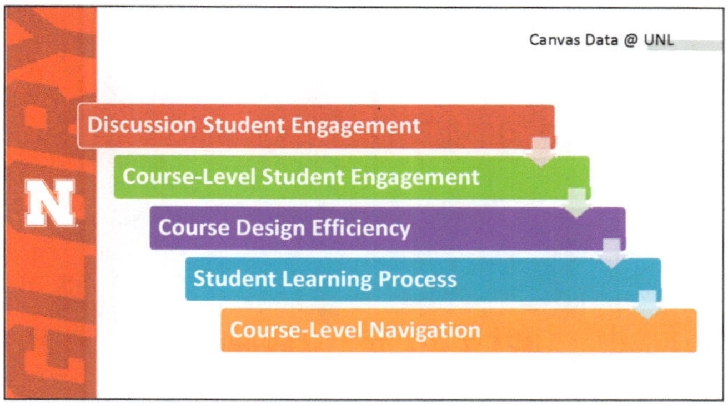

canvas

Slide 2

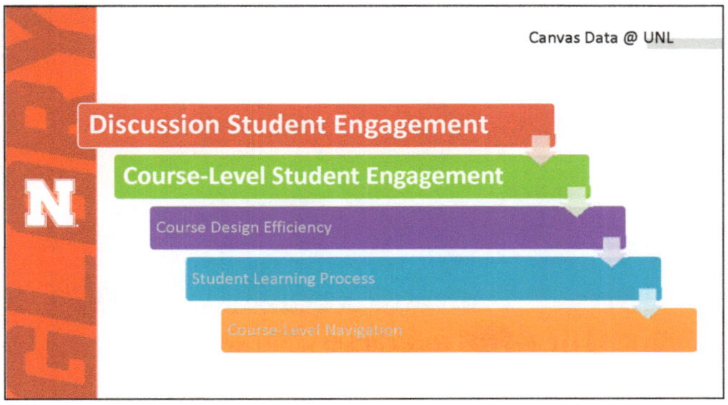

Canvas Data @ UNL

- Discussion Student Engagement
- Course-Level Student Engagement
- Course Design Efficiency
- Student Learning Process
- Course-Level Navigation

Slide 3

Canvas Data @ UNL

- Discussion Student Engagement
- Course-Level Student Engagement
- Course Design Efficiency
- Student Learning Process
- Course-Level Navigation

Slide 4

Student Engagement

Passive Engagement:
- Lowest barrier of student learning.
- Moving mouse and click the links.
- Exploring the course documents, pages, etc.

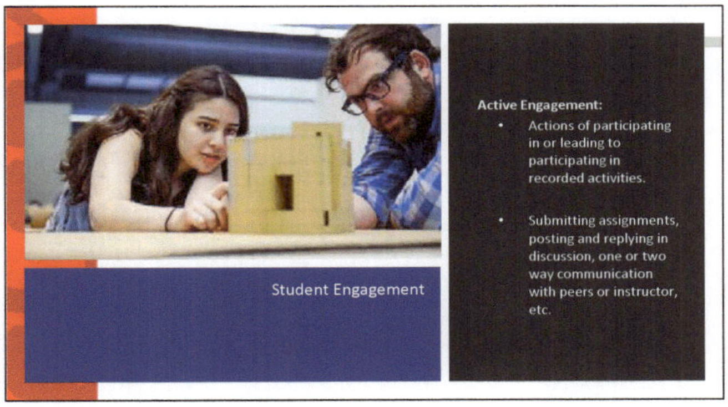

Slide 5

Student Engagement

Active Engagement:
- Actions of participating in or leading to participating in recorded activities.
- Submitting assignments, posting and replying in discussion, one or two way communication with peers or instructor, etc.

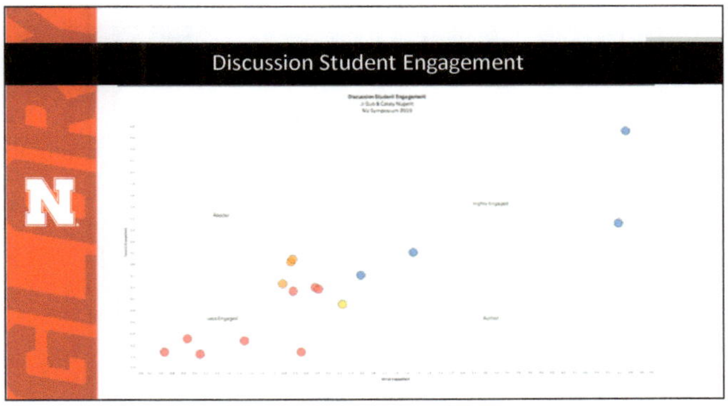

Slide 6

Discussion Student Engagement

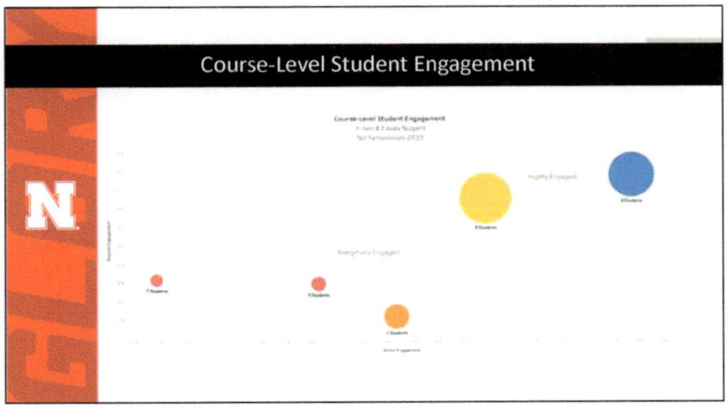

Slide 7

Course-Level Student Engagement

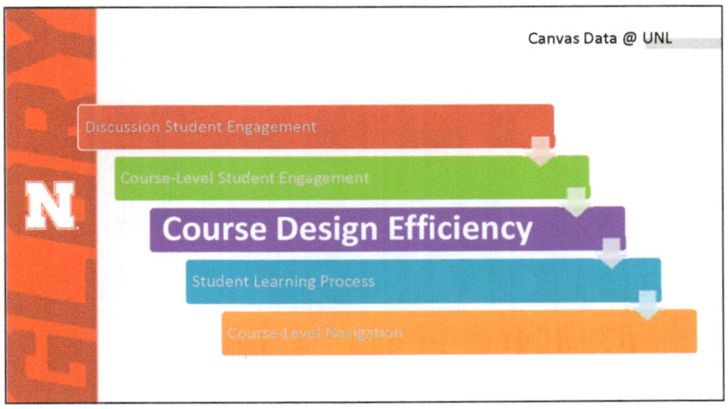

Slide 8

Canvas Data @ UNL

- Discussion Student Engagement
- Course-Level Student Engagement
- Course Design Efficiency
- Student Learning Process
- Course-Level Navigation

Course Design Efficiency Models

Course design efficiency models describe how much student effort converts to engagement and academic performance.

The thicker the arrow, the higher conversion efficiency.

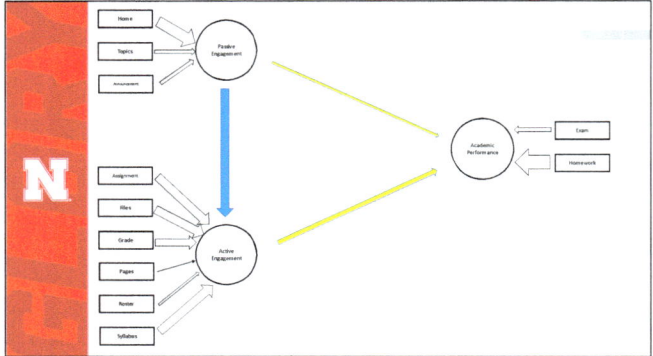

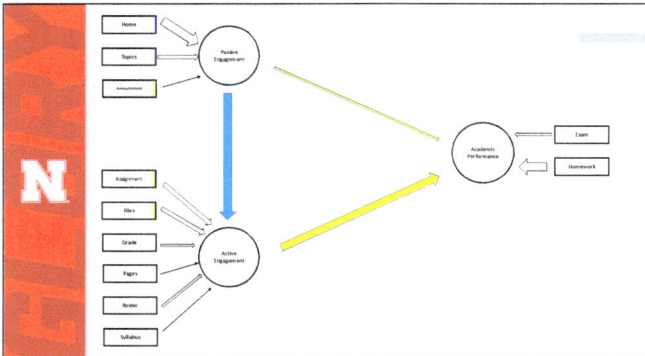

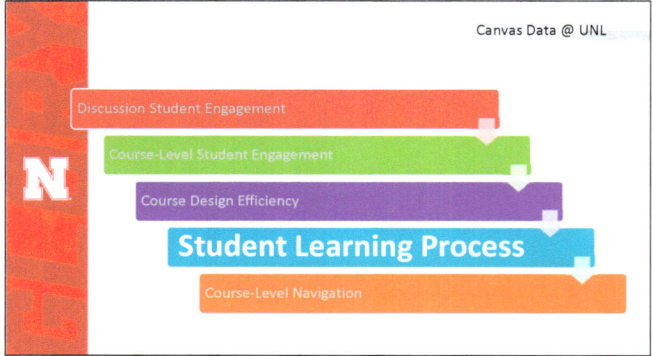

Canvas Data @ UNL

Discussion Student Engagement

Course-Level Student Engagement

Course Design Efficiency

Student Learning Process

Course-Level Navigation

Student Learning Process

Student learning process displays the most common behaviors of students browsing the course.

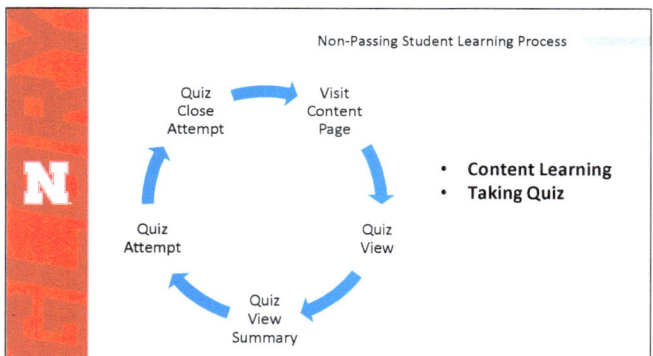

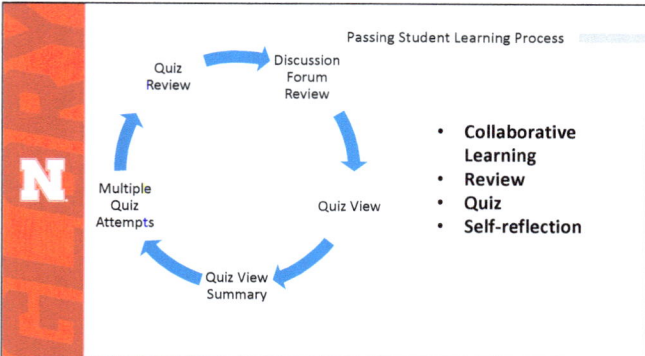

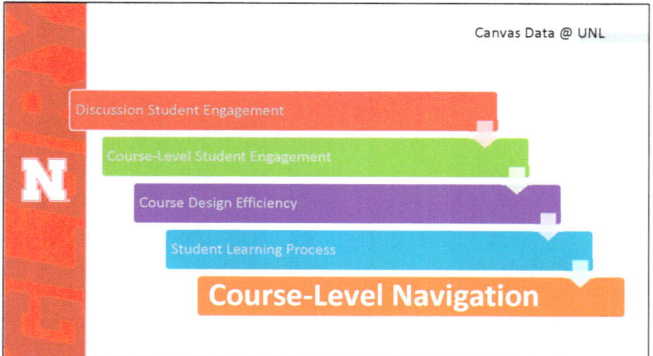

Non-Passing Student Learning Process

Quiz Close Attempt → Visit Content Page → Quiz View → Quiz View Summary → Quiz Attempt → Quiz Close Attempt

- **Content Learning**
- **Taking Quiz**

Passing Student Learning Process

Quiz Review → Discussion Forum Review → Quiz View → Quiz View Summary → Multiple Quiz Attempts → Quiz Review

- **Collaborative Learning**
- **Review**
- **Quiz**
- **Self-reflection**

Canvas Data @ UNL

Discussion Student Engagement

Course-Level Student Engagement

Course Design Efficiency

Student Learning Process

Course-Level Navigation

Student navigation patterns: Compare course designs

- Student navigation pattern diagram visualizes the inter-relationships between students and content categories. The connections display how frequently students access one content category.

- The wider the connections, the more student access.

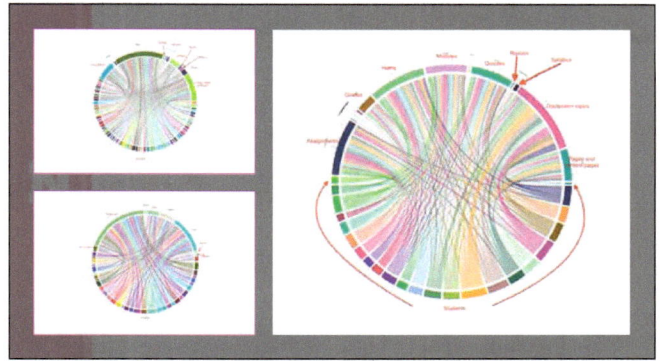

Future Plans @ UNL

- Integrate Data

- Use data to support and improve teaching and learning

- Create models to describe UNL student behaviors and actions

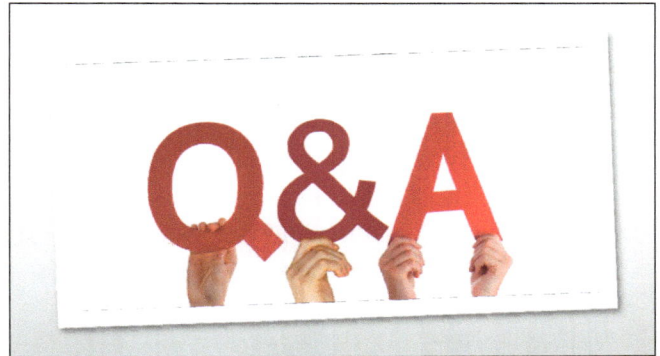

Contact Information

Ji Guo
IT Academic Technologies
ji@unl.edu
402-472-4456

Casey Nugent
Assistant Director
IT Academic Technologies
casey.nugent@nebraska.edu
402-472-5698

For more information, please visit:
Canvas Data and Student Learning Analytics
https://canvas.unl.edu/courses/27843

Midterm Evaluations: Making Midterm Course Corrections Using Meaningful Data

Ryan Caldwell (UNL)
Ben Lass (UNCA)
Tawnya Means, Ph.D. (UNL)
David Woodman (UNL)

End-of-term course evaluations have been the staple in higher education for years, with uses ranging from supporting promotion and tenure applications, selecting teaching awards and identifying curriculum and technology areas for improvement. However, these evaluations are limited in how they can support changes in real time for students. When instructors are able to get data in the middle of the semester, they have an opportunity to make meaningful instructional course corrections while they still have the students in class. These mid-course evaluations also let students know that their voice is being heard, thus making them more willing to provide feedback in the future. This presentation will take you through a journey of the meaningful value of midterm evaluations, the online and data analysis technologies used to make them happen, the potential for revealing valuable insights and strategies for responding to identified issues.

This presentation featured:

- the potential value of gathering feedback from students during the term and end of the term.
- discussion about the need for students to be actively involved in the feedback loop for courses.
- meta-analysis of student perceptions of teaching at the mid-point of a course that can be helpful in making corrections to improve the learning experience.

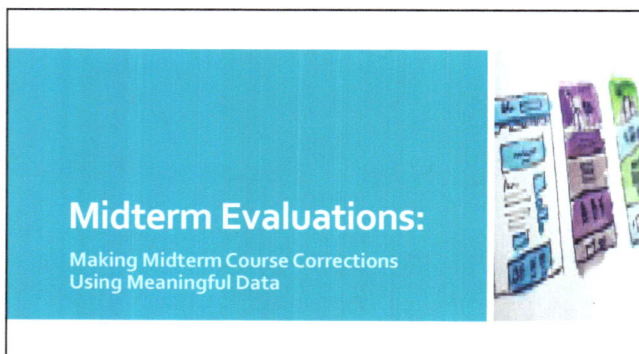

Midterm Evaluations:
Making Midterm Course Corrections Using Meaningful Data

We are:

- Tawnya Means
  Assistant Dean & Assistant Professor of Practice Management, UNL College of Business
- Ben Lass
  Manager, UNL Scanning Services
- Ryan Caldwell
  Assistant Director, UNL Institutional Effectiveness & Analytics
- David Woodman
  Professor of Practice, UNL School of Biological Sciences

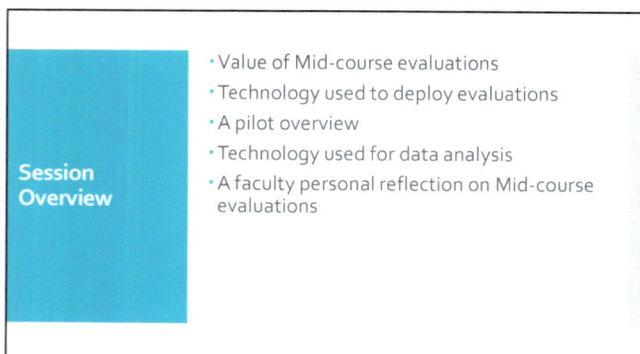

Session Overview

- Value of Mid-course evaluations
- Technology used to deploy evaluations
- A pilot overview
- Technology used for data analysis
- A faculty personal reflection on Mid-course evaluations

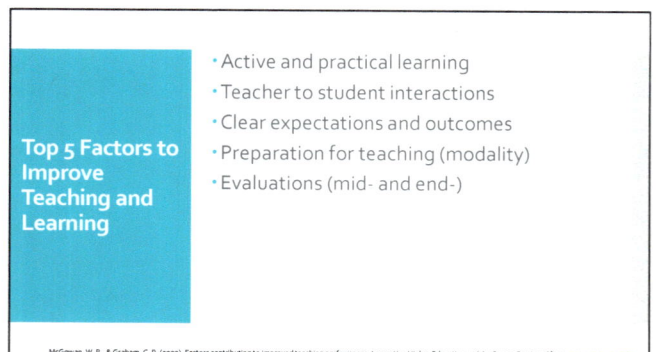

Top 5 Factors to Improve Teaching and Learning

- Active and practical learning
- Teacher to student interactions
- Clear expectations and outcomes
- Preparation for teaching (modality)
- Evaluations (mid- and end-)

McGowan, W. R., & Graham, C. R. (2009). Factors contributing to improved teaching performance. Innovative Higher Education, 34(3), 161-171. Retrieved from

Common Issues that Impact Evaluations

- **Course**: structure, design, logistics
- **Instructor**: Course communication, instructor presence
- **Tools**: Clarity and ease of use
- **Grading**: Student perception of weight of grades, fairness, response time
- **Assignments**: Instruction details, feedback
- **Teams**: Coordination, team dynamics, conflict management, purpose

The Value of Mid-course Evaluations

- "Check the pulse" of the class
- Show students you listen and care
- Evaluate student connection with learning (are they getting what you want them to get?)
- Opportunity for early intervention, course corrections and enhancements
- Evidence of commitment to teaching

Considerations for Mid-course Evaluations

- Timing: After first assessment, but early enough to make changes
- Closed vs. Open ended questions (depends on scale of the course)

- What do you want to know?
- What will you do with the input?

How to conduct Mid-course Evaluations

- EvaluationKit provides for anonymous collection of student responses
- Integrates with Canvas gradebook (completion points)
- Explain to students why you want feedback ([tips on how to provide useful feedback](#))
- Evaluate data results
  - What can I address right **now**?
  - What can I address in a **future** term?
  - What **can't be changed** (but I can better explain why I do what I do)?

- Give feedback that avoids "emotionally-charged" words
- Give feedback that describes specific behavior rather than your inferences
- Give feedback that reflects on positive behaviors and gives solutions

Less helpful comments	Helpful comments
This professor was awesome.	This professor gave us lots of activities to do in the classroom which helped me REALLY understand what I was doing rather than remembering stuff for a quiz.
This professor sucks.	I had a problem understanding all the jargon in many of the lectures. It would have helped me if you talked in plainer English.
This professor doesn't like students.	I wish you had talked to the class like you were teaching people and not just the content of his subject. I would have liked you to check for understanding at certain points in the lecture.
The professor just talked at us in the lecture.	I would have preferred if we could have done some group work to discuss some of the ideas with our peers.
This professor was motivating.	This professor told us stories about how he collected data for his research and made me enthusiastic to want to do the same.
The professor never went over the homework.	It would have been helpful if the professor incorporated a few of the homework problems into the lecture so we could have seen how to go about solving them and could ask questions if we still didn't understand.
The professor was caring.	I really appreciated the way the professor was always there after class to answer questions and always responded to my e-mails.

https://facultyinnovate.utexas.edu/sites/default/files/giving_useful_feedback_to_your_professors_12716.pdf

---

EvaluationKit as a tool

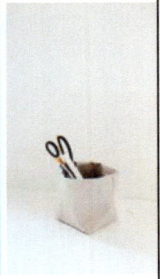

What is EvaluationKit?

- On-line course evaluation & survey software
- Multiple roles
  - Administrator, Sub-administrator, Instructor, TA, Student
- Integrated w/ LMS
  - Deployed in Canvas
  - Pulls data from Canvas, not SIS

What is EvaluationKit?

- Robust reporting features
  - Faculty, department, college, institution levels
  - Cross-tabulation and longitudinal reporting
- Multiple notification and reminder features to increase response rates
- Multi-level support---phone, e-mail, video tutorials, help articles

---

Student Features

- Dashboard on course page menu
- Pop-up notifications
- E-mail reminders
- Can add a Help Button
- "Opt-out" option

Student Dashboard

Pop-Up Notifications

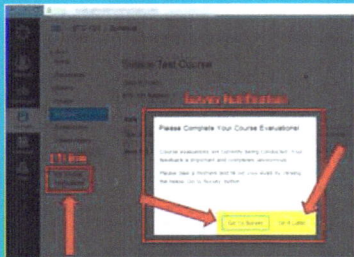

Opt Out Feature

Faculty Features

- Dashboard on course page menu
- Response Rate Tracker
- Access to results
- Can create and add questions and manage courses

Faculty View

Pilot Description & Scope

12 questions
72 instructors
130 courses

Pilot Description & Scope

10 days
(2/11 – 2/20/2019)
5435 survey enrollments
4371 unique users (students)

Pilot Results

74.98%
response rate
(4074/5435 survey enrollments)

Pilot Results

- What worked well:
  - Notifications/Reminders
  - Response Rate Trackers for faculty
  - Open ended questions – students wrote a lot!

Pilot Results

- Challenges:
  - Notifications/Reminders--too invasive and overwhelming (until we fixed them)
  - Discovering situations we were unaware that existed
    - Lab courses
    - On-line exams in classroom setting

Instructor Post-Pilot Survey & Results

9 questions
72 instructors
130 courses

Instructor Post-Pilot Survey & Results

60.77%
response rate
(79/130 courses)

Instructor Post-Pilot Survey & Results

- What worked well:
  - Happy with results
  - Gained useful feedback
  - Plan to make course adjustments
  - Would use the survey again

Instructor Post-Pilot Survey & Results

- Challenges
  - Questions not specific enough for their particular course
  - Format of questions needs work
  - Evaluation too early in the semester
  - Pop-up notifications and reminders were troublesome
  - Prefer a 5-point Likert scale

Visualizing the Data

Automated Reporting through Tableau

- What is Tableau?
  - Data Visualization
  - Report Automation
  - Granular Security
- Organize the data quickly so the instructor can spend time analyzing results rather than compiling them
- Pilot Results

One Instructor's View

Evaluation Timing

- Determined by the instructor (flexible)
  - After students have experienced the most significant feedback components that a class offers
  - Mid-semester
- Early enough to permit change

Question Types

- Same as in end-of-semester survey (university-wide questions)
  - Too early to make a judgement
  - Fatigue
- College or Department determined questions
  - Narrower in focus
- Instructor determined questions
  - Narrowest focus, pertinent to a few foci
  - Most relevant to the goal of making a change

Preferred Format

- Did lab "X" facilitate your learning?
- Were the lab and lecture well coordinated?
- Were the TA's well trained?
- Did you use the video podcast? How often and when?
- What is your opinion of the online tools (video, ppts, podcasts) used for Lab/lecture?

Interpretation Large vs. Small Classes

Desirable Features

6 - My attendance in this class so far:

Response Option	Weight	Frequency	Percent	Percent Responses	Means
missed less than 3 classes	(1)	164	99.39%		
missed 3 - 5 classes	(2)	1	0.61%		
missed one fourth of classes	(3)	0	0.00%		
missed more than half of classes	(4)	0	0.00%		

Response Rate
165/209 (78.95%)

7 - My anticipated grade for this class is:

Response Option	Weight	Frequency	Percent	Percent Responses	Means
A	(1)	42	25.45%		
B	(2)	82	49.70%		
C	(3)	21	12.73%		
D	(4)	3	1.82%		
F	(5)	0	0.00%		
Do not know	(6)	17	10.30%		

Response Rate
165/209 (78.95%)

Interpretational Problems

3 - 2. What's been beneficial to your learning? From the following list of teaching elements, what is the one thing that has been the most beneficial for your learning this term so far? Please provide written comments about the element you selected.

David Woodman

Response Option	Weight	Frequency	Percent	Percent Responses	Means
Inclusiveness	(1)	1	0.63%		
Transparency of instructions and grading	(2)	5	3.14%		
Timing of feedback	(3)	5	3.14%		
Challenge of the course	(4)	40	25.16%		
Quality and accessibility of course materials	(5)	46	28.93%		
Support	(6)	3	1.89%		
Engagement in assignments or projects	(7)	4	2.52%		
Active Learning	(8)	22	13.84%		
Quality of the interactions between students	(9)	13	8.18%		
Instructor communication	(10)	7	4.40%		
Other	(11)	3	1.89%		
Not applicable	(0)	10	6.29%		

Response Rate
159/209 (76.08%)

4 - 3. What could use some improvement? From the following list of teaching elements, what is the one thing that could most use some improvement to increase student learning? Please provide written comments about the element you selected.

David Woodman

Response Option	Weight	Frequency	Percent	Percent Responses	Means
Inclusiveness	(1)	4	2.58%		
Transparency of instructions and grading	(2)	13	8.39%		
Timing of feedback	(3)	2	1.29%		
Challenge of the course	(4)	25	16.13%		
Quality and accessibility of course materials	(5)	9	5.81%		
Support	(6)	12	7.74%		
Engagement in assignments or projects	(7)	7	4.52%		
Active Learning	(8)	16	10.32%		
Quality of the interactions between students	(9)	10	6.45%		
Instructor communication	(10)	7	4.52%		
Other	(11)	5	3.23%		
Not applicable	(0)	45	29.03%		

Response Rate
155/209 (74.16%)

Comments linked to elements selected, but not connected to them in presentation of results

Question	3. What could use some improvement? From the following list of teaching elements, what is the one thing that could most use some improvement to increase student learning? Please provide written comments about the element you selected : David Woodman	#	%
1	Inclusiveness	4	2.58%
2	Transparency of instructions and grading	13	8.39%
3	Timing of feedback	2	1.29%
4	Challenge of the course	25	16.13%
5	Quality and accessibility of course materials	9	5.81%
6	Support	12	7.74%
7	Engagement in assignments or projects	7	4.52%
8	Active Learning	16	10.32%
9	Quality of the interactions between students	10	6.45%
10	Instructor communication	7	4.52%
11	Other	5	3.23%
0	Not applicable	45	29.03%

Formative Data difficult to interpret

Long-term data

Who should see it? Privacy issue

- Faculty member only
  - Perhaps I am taking a risky approach and I don't want this to be part of a record
  - I have very specific content areas I need evaluated pedagogically and it has no bearing on my teaching as a whole
- Faculty member plus mentor
  - I am providing this information to a trusted mentor who shares my goal to improve my teaching without exposing my vulnerabilities
  - My mentor can help me analyze the information and provide guidance
  - My mentor can assist me to change my strategies if the one I am using does not work
- Faculty member plus Teaching Assistants

Questions?

Thank You!

Mindful Pause Practice: The How To's and Why To's of Adding Mindfulness to Your Course

Tanya Custer (UNMC)
Kim Michael (UNMC)

Mindfulness can be defined as the awareness that emerges through paying attention on purpose, in the moment and non-judgmentally to the unfolding of experiences moment by moment. Mindful interventions are shown to improve mental, emotional and physical well-being as well as cognition and academic performance. 25 nursing students who were exposed to a 3-5 minute mindful practice at the start of each class reported being more calm and relaxed, having better focus on the professor and class content, lowering of internal and external distractions and an increasing sense of community and connection to their classmates and instructor. In this presentation, participants will complete the Mindful Attention Awareness Scale to determine their mindful score, explore the benefits of adding a mindful intervention to a course, how to add a mindful intervention to a course and potential outcomes of this type of intervention.

This presentation featured:
- the Mindful Attention Awareness Scale (MAAS) score.
- how mindfulness can improve mental, emotional and physical well-being.
- mindful techniques which can be utilized within an online or face-to-face course.

Fostering Conversations with Faculty about Quality Online Courses

Kristin Bradley (UNCA)
Erin King (UNCA)

There are certain qualities and criteria that the industry agrees go into the making and composition of courses that are well designed and easy for students to use. The key is identifying what those are and getting faculty and instructional designers to ensure that they're incorporated. The Office of Digital Learning at the University of Nebraska Omaha has taken the OpenSUNY OSCQR Rubric and combined it with a Canvas Course Checklist to create a comprehensive Course Design Rubric. This rubric is provided to faculty who are developing new online courses or converting on-campus courses to online; as well as being used by instructional designers to review online courses at a faculty's request. The rubric is color-coded so that faculty can see at a glance which criteria are expected to be part of every course, which required more advanced skills and those that are considered expert level. This presentation will discuss how we are using the rubric to foster conversations with faculty about best practices for online course design.

This presentation featured:

- ways to start having the "quality" conversations with faculty.
- ideas for talking to faculty about the importance of accessibility in online courses.
- how a course quality rubric can be used on your campus.

Final Grades Integration for Efficiency

William Barrera (UNCA)
Marcia L. Dority Baker (UNCA)
Matthew Schill (UNO)
Tomm Roland (UNO)

The purpose of this project is to gain efficiencies when entering grades in both Canvas and PeopleSoft. This new integration will allow faculty to pull grades into PeopleSoft on a class by class basis. The integration will reduce the time that faculty spend entering grades in two systems. This presentation will review the Spring 2019 pilot at UNL including project success, learning outcomes and modifications for the integration to go live for the NU System beginning with the 2019-20 academic year.

This presentation featured:

- the relationship between PeopleSoft and Canvas.
- how to integrate two systems for efficiencies.
- how ITS pilots projects to improve workflow for campus stakeholders.

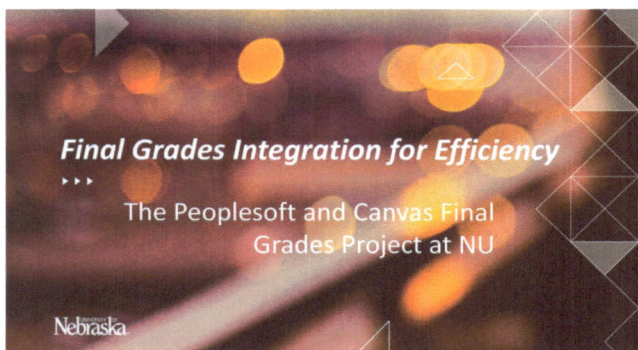

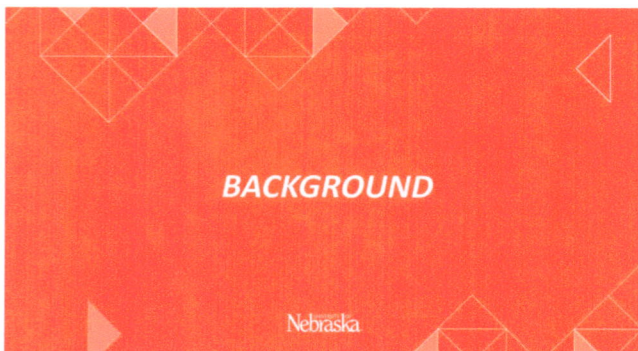

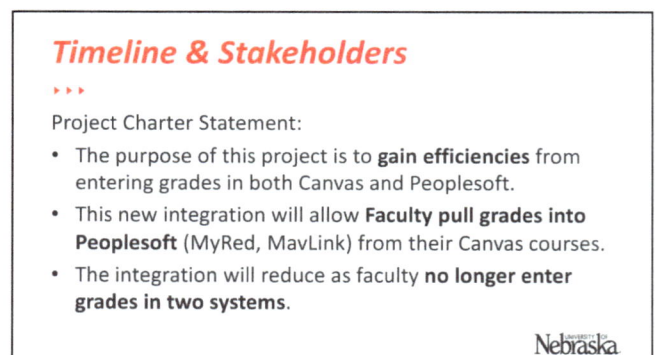

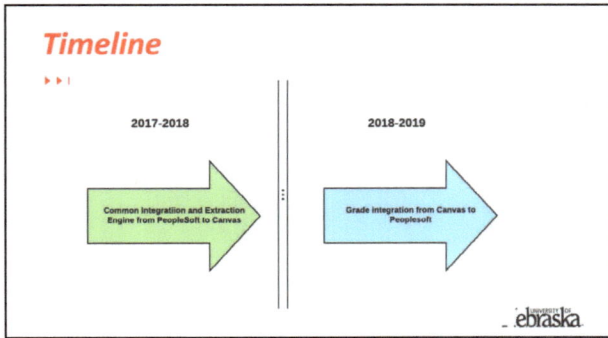

Timeline

▶ ▶ ▮

2017-2018

Common Integration and Extraction Engine from PeopleSoft to Canvas →

2018-2019

Grade Integration from Canvas to Peoplesoft →

University of Nebraska

Solicit Faculty Participation

- Communication, working groups, Faculty Senate
- Documentation and workflow
- Spring pilot at UNO and UNL

ONE UNIVERSITY. FOUR CAMPUSES. ONE NEBRASKA. Nebraska

Timeline

▶ ▶ ▶

- Communication with Pilot Faculty
  - Early February – identify Faculty
  - March – Formal invite from Registrar
  - Spring Break – Test training walkthrough
  - Late April – Grading Open!

University of Nebraska

What's It Look Like?

ONE UNIVERSITY. FOUR CAMPUSES. ONE NEBRASKA. Nebraska

Faculty feedback...

Qualtrix Survey launched April 29

Nebraska

Faculty feedback...

How much time did it save?

- 20-40 minutes
- More than 60 minutes
- 20-40 minutes

Nebraska

Faculty feedback...

General Comments?

- It seemed to work very smoothly
- This is awesome! It just worked.

Nebraska

Faculty feedback...

Suggestions

- Make sure folks understand difference in default grading scheme and their course grading scheme
- Need to recreate grading scheme for each course

Nebraska

What's Next

- Spring 2019 pilot and feedback, UNO and UNL
- Summer 2019 pilot and feedback, UNMC, UNK
- Fall 2019 go live across NU system

ONE UNIVERSITY. FOUR CAMPUSES. ONE NEBRASKA. Nebraska

Questions

ONE UNIVERSITY. FOUR CAMPUSES. ONE NEBRASKA. Nebraska

Small Change, Big Impact: Bringing Active Learning to the Online Environment

Grace Troupe (UNL)

We're all looking for ways to reach through the screen to better engage our online students. This presentation will demonstrate four methods to use in classrooms in order to bring active learning to the online environment: learning quizzes, muddiest point, Zoom labs and two-phase collaborative exams. All four methods are derived from evidence-based techniques. The methods are inexpensive and easy to incorporate into your course, yet can have a big impact on learning and online engagement.

This presentation featured:

- short lectures and quizzes to break the fluency illusion and close the feedback loop.
- closing the feedback loop for both students and teachers thereby modeling optimal communication.
- Zoom labs to bring real-time small group problem solving to an online class.
- the two-phase, collaborative, individual/group exams to complement interactive teaching methods and improve learning outcomes.

While you're waiting, **talk with your neighbor** about:

- What **you teach**
- Your **course set-up**
  - Large lecture?
  - Online?
  - Small class?
- What **methods** you use to deliver content and engage learners

We'll save **time to discuss** the methods shared and how they could be useful in your learning environment.

N

Grace Troupe works from home as a part-time **online instructor**.

N

Small Change, Big Impact:
Bringing Active Learning to the Online Environment

Grace Troupe
Online Instructor
gtroupe2@unl.edu

N

The University of Nebraska does not discriminate based upon any protected status. Please see go.unl.edu/nondiscrimination.

I will share **four methods** I use to bring **active learning** to the **online classroom**.

Muddiest Point

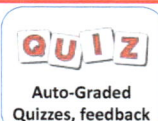
Auto-Graded Quizzes, feedback

Zoom Labs

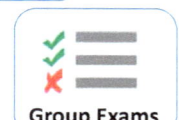
Group Exams

N

What do I mean by **active learning**?

Any teaching method targeted at student involvement that requires the learner to go beyond listening.

Flipped classroom

THINK (Yourself)

PAIR (With a partner)

SHARE (Whole class)

The **learning cycle** follows the **same structure** each week in both the online and face-to-face course.

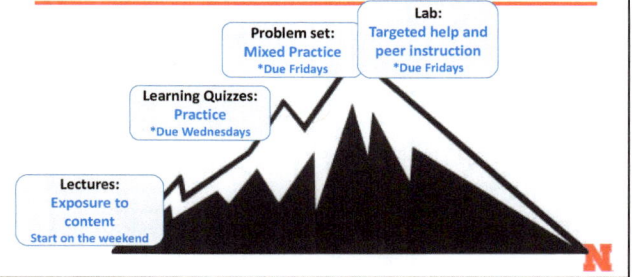

Problem set: Mixed Practice *Due Fridays

Lab: Targeted help and peer instruction *Due Fridays

Learning Quizzes: Practice *Due Wednesdays

Lectures: Exposure to content Start on the weekend

The **learning cycle** repeats for 3 weeks/lessons to make a unit, culminating with an exam.

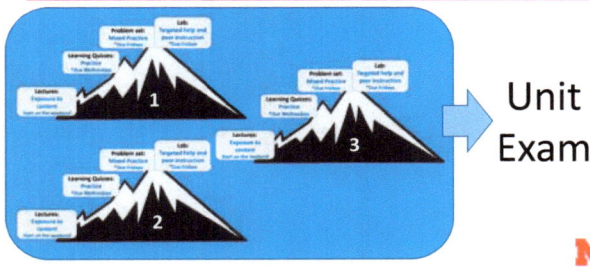

1 2 3

Unit Exam

Use **muddiest point** to gain insight to your student's learning and provide targeted instruction.

Muddiest Point

Muddiest point encourages **online students** to **share** what they are **struggling** with.

Muddiest Point

Screen shot of their hardest quiz question

One sentence about why this question was muddy for them.

Students report the **least clear idea** from the class period.

Muddiest point provides **broad, asynchronous** help to students.

Muddiest Point

Student

- **Models communication** with online instructor
- Responsive to **student needs**
- Incentivizes **reflection** and question asking for **ALL** students

Instructor

- **Condenses feedback** to make **whole-class interaction** realistic
- **Reduces e-mail** load
- Creates **connection** with students
- Feedback to the instructor about **what is 'working'**
  Carberry, A., Krause, S., Ankeny, C., & Waters, C. (2013, October). "Unmuddying" course content using muddiest point reflections. In *2013 IEEE Frontiers in Education Conference (FIE)* (pp. 937-942). IEEE.
- **Reusable** semester to semester, can become proactive help videos linked with the quizzes

Benefits

Share with your neighbor:

How many times a week do students ask you a **content** question?

Use **quizzes** to bust the **fluency illusion** and give timely **feedback**.

QUIZ
Auto-Graded Quizzes, feedback

What is the **fluency illusion**?
Lecture / Group Work — **Fluency Demonstrated**
BY GEORGE / I THINK I'VE GOT IT
Fluency mistaken as your own

Quizzes can **bust the fluency illusion.**
Bite-sized Lecture / Group Work — **Fluency Demonstrated**
QUIZ
BY GEORGE / I THINK I'VE GOT IT
Fluency mistaken as your own

Feedback in the learning moment debunks hard-to-correct misconceptions.

Quizzes fill the
Cancel / Update Question
Canvas allows you to add feedback that will show after the quiz is submitted to hit in the 'learning moment.'

Auto-graded quizzes give instant feedback.

Benefits

Student
- Asynchronous **feedback** in the **learning moment**
- Promotes **self-driven learning** and reflection
- Ideal for the **online** student

Instructor
- **Reduced** grading **time**
- Once it is set up, it is '**hands off**' help
- **Flexible use** for in-person and online audiences

Zoom Labs
Use **Zoom labs** to add face-to-face peer instruction to an online course.

Zoom allows for the **online equivalent** of an on-campus recitation.
On-campus recitation → Zoom labs

In an online lab, a '**breakout room**' is a small table group where you engage in **team problem solving**.
Table group = Breakout room

Breakout rooms must be 'turned on' in the Zoom advanced settings in your account at Zoom.us.

Zoom Labs

In Meeting (Advanced)

Breakout room
Allow host to split meeting participants into separate, smaller rooms

---

How can I add peer instruction to an online class?

Zoom Labs

Groups **can call in the teacher** for help and oral questions

Shared screen with simultaneous **annotation** for team problem solving

Breakout rooms allow for **small groups**

Zoom Labs

What is it actually like in a Zoom lab?

Here is a video that includes clips of me working with students in my Zoom labs: https://www.youtube.com/watch?v=k-u55w6Z_SE

Zoom Labs

Both students and instructors **benefit** from **Zoom** labs.

Student
- Consistent **access to help** from peers and an instructor/TA
- **Normalize** question **asking**
- **Targeted teaching** on misconceptions
- **Real-time** feedback
- Real **connection** with other learners

Benefits

Instructor
- **Reduced** grading **time**
- Student's **responses to the content** are more apparent
- **Easy** to implement, **free** (UNL has a license)

Group Exams

Use **peer instruction** to squeeze extra learning out of your exam.

Testing
Group Exams

A group exam has two phases.

Testing

Phase 1:
Individual exam (90%)

Testing

Phase 2: Take the **same exam again with your team (10%)**

Take advantage of the **busted fluency illusions** with an added **group exam**.

Testing
Group Exams

Group exams are a **research supported** strategy that benefit both student and instructor.

Student
- **Fresh awareness** of what students DO NOT know **fuels peer instruction**
- Students **love it**
- Rewards **peer instruction**
- **Builds confidence**

Benefits

Instructor
- **Students** are more **motivated** in an exam setting
- Very **easy** to implement

Gilley, B. H., & Clarkston, B. (2014). Collaborative testing: Evidence of learning in a controlled in-class study of undergraduate students. Journal of College Science Teaching, 43(3), 83-91.

---

3 min: Think about how these methods **apply to your learning environment**.

Muddiest Point — Use **muddiest point** to gain insight to your student's learning and provide targeted instruction.

Auto-Graded Quizzes, feedback — Use **quizzes** to bust the **fluency illusion** and give timely **feedback**.

Zoom Labs — Use **Zoom labs** to add face-to-face peer instruction to an online course.

Group Exams — Use **peer instruction** to squeeze extra learning out of your exam.

Talk with your neighbor about how these could be adapted for your audience and **share with the group**: https://go.unl.edu/r9tq

Muddiest Point

Auto-Graded Quizzes, feedback

Zoom Labs

Group Exams

Increase Online Class Size & Student Satisfaction Without Increasing Faculty Workload

B. Jean Mandernach, Ph.D. (UNK)
Steven McGahan (UNK)

Student demand for access to online classes often exceeds course availability or traditional class size limits. While there is considerable literature outlining strategies for increasing class size by automating learning activities via technology or utilizing teaching assistants, there is a dearth of research exploring strategies to increase online class size while retaining high levels of personalized interaction/feedback with a single instructor. The presentation overviews online course design strategies that allow an instructor to double (or triple) course size while maintaining opportunities for frequent student-instructor interaction, individualized feedback and a personalized learning experience. We will provide a comparative analysis of student learning outcomes, satisfaction and engagement from students enrolled in either a 25- or 60-person online general psychology course.

This presentation featured:
- strategies for increasing course size without increasing faculty load.
- opportunities to increase student satisfaction via choice of learning activities.
- tips for creating an interactive, personalized learning experience that doesn't overwhelm instructional time.

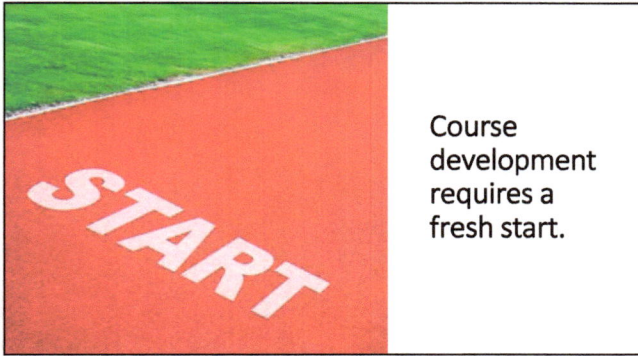

Course development requires a fresh start.

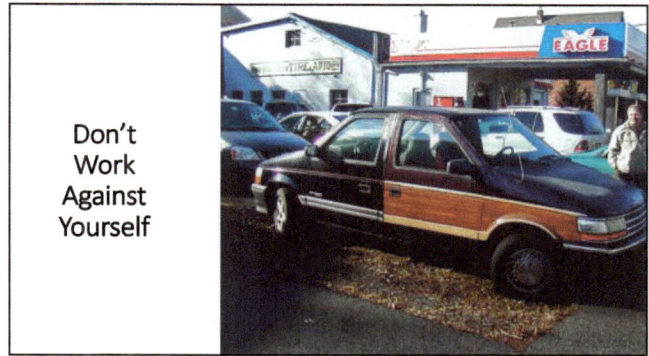

Don't Work Against Yourself

Practical Considerations

- Faculty availability
- Course development
- Quality control
- Support
- Time
- Money

How do we effectively educate more students in a single online course?

The bad...

(Ziker, J. (2014). The long, lonely job of homo academicus, *The Blue Review*. https://thebluereview.org/faculty-time-allocation/)

61 hours per week

12%	12%	11%	35%
Instruction	Class Preparation	Course Administration & Grading	**TEACHING**

A little math...

21.35 hours per week

- 3 courses → 7.12 hours per course per week
- 4 courses → 5.34 hours per course per week

...and more math...

7.12 hours

5.34 hours

6.23 hours per class per week for campus faculty

ONLINE Teaching

Adjunct | Fulltime

13.33 hours | **11.05 hours**

per course per week

11.63 (online)

+

6.23 (campus)

≠

9 hours per week per online class

The ugly...

If you have 9 hours per week...

25 students	50 students	75 students
• 21.6 minutes	• 10.8 minutes	• 7.2 minutes
• per student	• per student	• per student
• per week	• per week	• per week

Available Time **=** Instructional Tasks

- Course development
- Technical challenges
- Course interaction
- Grading & feedback
- Communication
- Course administration
- Content development

To maximize instructional effectiveness...

Prioritize teaching time to focus on high impact instructional activities

interaction ➡ presence ➡ feedback

Think back to your Undergraduate Degree

One thing

Prioritize the Important Content

The Challenge

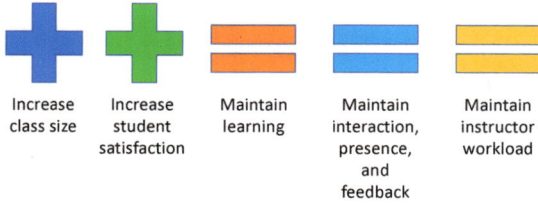

Increase class size	Increase student satisfaction	Maintain learning	Maintain interaction, presence, and feedback	Maintain instructor workload

Context: General Psychology

Course Context:
- General studies
- Mainly non-majors
- Mix of traditional and non-traditional students
- 16-week schedule
- Fully online

Learning Objectives:
- Introduce students to the diverse field of psychology.
- Demonstrate that psychology is a research discipline with important human applications.
- Teach students to become intelligent consumers of psychological research.
- Familiarize students with current trends in technology that influence the field of psychology.

Where do "good" online faculty spend their teaching time?

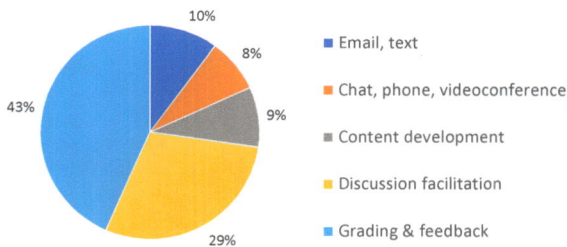

- 10% — Email, text
- 8% — Chat, phone, videoconference
- 9% — Content development
- 29% — Discussion facilitation
- 43% — Grading & feedback

Assignment Structure

Required of all students:
- Weekly Mastery Quiz
- Final Exam Review
- Final Exam

Choice of one of the following per week:
- Discussion
- Journal
- Research Analysis
- Video Exploration
- Current Event Analysis

(Assignment Descriptions)

Discussion — Online discussions explore the application and analysis of psychological concepts. You must post your response to the initial discussion question and a minimum of 4 peer replies.

Journal — The journal is your opportunity to apply course concepts in an analysis of the world around you. Journal entries should be 250-500 words (12-point font, 1-inch margins, double-spaced; do NOT include a heading or your name at the top of the journal entry).

Research Analysis — The research analysis requires you to read a selected journal article and analyze its value/relevance. Each research analysis includes 20 multiple-choice questions and 1 essay designed to test your ability to understand, critically evaluate and apply information from psychological research.

Video Exploration — The video exploration is your opportunity to investigate ONE selected course topic in more detail. Your task is to learn more about the topic than what is presented in our textbook. You will then create a video of yourself sharing the information you found. Your video must be between 1- and 2-minutes. You do not need any graphics or other multimedia; the video should be of you talking into the camera.

Current Event Analysis — The current event analysis is designed to allow you to further explore psychology in the real world. Submit an identification of the psychological concepts relevant to your current event and an explanation of how the psychological concepts apply. Utilize a bulleted list in which you identify each concept then include a few sentences in which you explain the relationship.

Skills or Concepts?

Distribution of Workload

- Course is divided into 3 blocks; each block contain 5 weeks/chapters
- Students select and complete 1 of each assignment type per block
- At the end of the course, each student has completed 3 of each assignment type (for a total of 15 assignments)

Assignment Plan

Block	Week	Topic	Mastery Quiz	Final Exam	Journal	Discussion	Research Analysis	Video Exploration	Current Event Analysis
1	1	Course Overview	X						
	2	Introduction to Psychology	X						
	3	Research Methods	X						
	4	Brain & Nervous System	X						
	5	Sensation and Perception	X						
2	6	States of Consciousness	X						
	7	Human Development	X						
	8	Learning	X						
	9	Memory & Decision-Making	X						
	10	Intelligence & Language	X						
3	11	Emotions, Motivation & Stress	X						
	12	Social Psychology	X						
	13	Personality	X						
	14	Mental Health	X						
	15	Therapy	X						
	16	Review & Final Exam	X	X					

Grading Structure

Assignment	Points
Final Exam (140 points)	140
Final Exam Review (20 points)	20
Mastery Quizzes (10 points each)	150
Discussion (30 points each)	90
Journal (20 points each)	60
Research Analysis (25 points each)	75
Video Exploration (30 points each)	90
Current Event Analysis (25 points each)	75
TOTAL	700

Instructor Workload

- 5 day work week
- 60 students (12 students per assignment type)

Assignment	Ongoing Time Investment (per day)	Grading & Feedback Time (end of week)	Approximate Grading Time Per Student	Weekly Discussion Interaction	Weekly Grading Time
Discussion	30 minutes	24 minutes	2 minutes	150 minutes	24 minutes
Journal	0 minutes	72 minutes	6 minutes	0 minutes	72 minutes
Research Analysis	0 minutes	12 minutes	1 minute	0 minutes	12 minutes
Video Exploration	0 minutes	36 minutes	3 minutes	0 minutes	36 minutes
Current Event Analysis	0 minutes	54 minutes	4.5 minutes	0 minutes	54 minutes
Mastery Quiz	0 minutes	0 minutes	0 minutes	0 minutes	0 minutes
TOTAL				150 minutes	198 minutes

"Best practice" if you have 9 hours per week to spend on your online teaching...

	Time Allocation	Time Per Week	Instructor Workload with Increased Class Size
Email, text	10%	54 minutes	86 minutes
Chat, phone, videoconference	8%	43.2 minutes	20 minutes
Content development	9%	48.6 minutes	30 minutes
Discussion facilitation	29%	156.6 minutes (2.61 hours)	150 minutes
Grading & feedback	43%	232.2 minutes (3.87 hours)	198 minutes
			484 minutes (8.07 hours)

The outcome...

Increased class size + Increased student satisfaction = Equivalent learning + Increased perception of interaction and presence − Decrease instructor workload

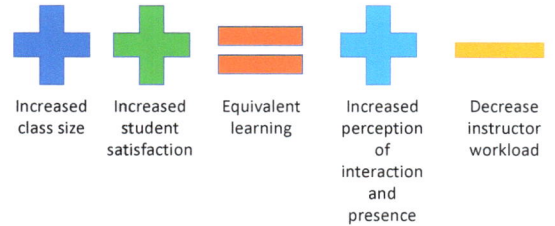

Student feedback...

Choice in assignments
Variability
High levels of interaction

Gradebook confusion
Uncertainty of LMS directions

Discussion	Journal	Research Analysis	Video Exploration	Current Event Analysis	Mastery Quiz
• Loved the interaction • Liked not doing it every week • Found questions interesting	• Found value in the personal application • Liked relating to major and/or career interests	• Hated it • Found it irrelevant to course • Thought it was very difficult	• Mixed reviews (most loved it) • Liked the novelty • Enjoyed exploring personal interests	• Considerable confusion over what is considered a current event • Struggled to apply to course concepts	• Believed useful but boring • Liked being able to take the quiz multiple times for better grade

Questions or comments?

Email Deception & Trickery

Cheryl O'Dell (UNCA)
Nick Glade (UNCA)
JR Noble (UNCA)

Phishing continues to be a top cyberattack experienced by our institution. According to the annual Verizon Data Breach Investigation Report, phishing has contributed to numerous data compromises globally since 2008. There has been and continues to be a tremendous effort to slow down phishing attacks, but ultimately, the way to fight phishing is for an effort to be made by every employee that uses email. Come hear about why phishing continues to be a successful threat vector, common phishing schemes (some more successful than others), ways to recognize a phish email, ways that the security and email support staff are fighting phishing and what to do if you suspect you have received a phish email.

This presentation featured:

- tips on how to spot email trickery on your mobile device or computer.
- the NEW way to report suspicious emails.
- a how-to guide for email security solutions deployed at each campus.

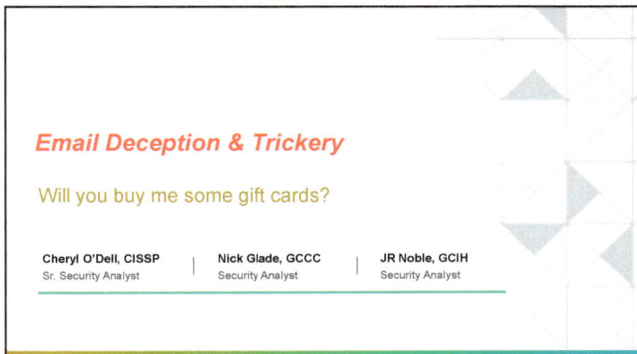

Email Deception & Trickery

Will you buy me some gift cards?

| Cheryl O'Dell, CISSP | Nick Glade, GCCC | JR Noble, GCIH |
| Sr. Security Analyst | Security Analyst | Security Analyst |

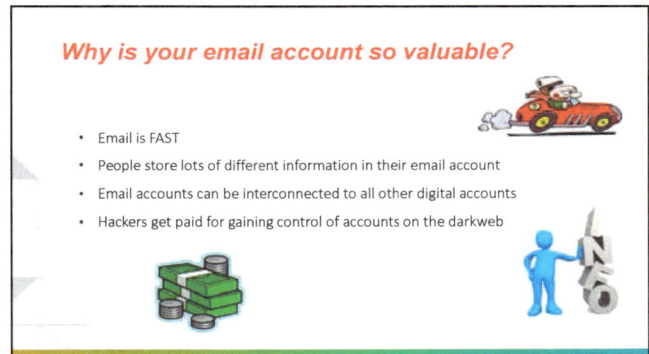

Why is your email account so valuable?

- Email is FAST
- People store lots of different information in their email account
- Email accounts can be interconnected to all other digital accounts
- Hackers get paid for gaining control of accounts on the darkweb

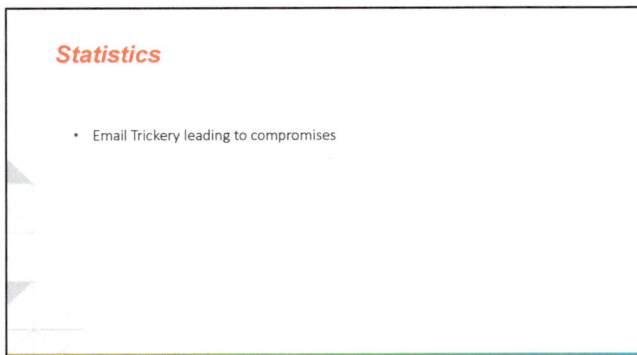

Statistics

- Email Trickery leading to compromises

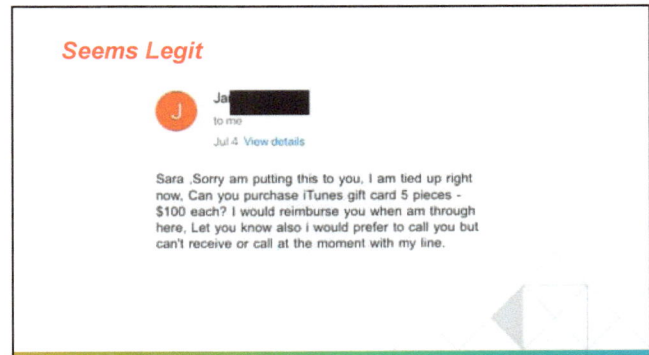

Seems Legit

Sara ,Sorry am putting this to you, I am tied up right now, Can you purchase iTunes gift card 5 pieces - $100 each? I would reimburse you when am through here, Let you know also i would prefer to call you but can't receive or call at the moment with my line.

Business Email Compromise

- Hacker creates fake Gmail or Yahoo account using CEO's name
- Targeted at staff with financial access
- Often involve purchasing gift cards or transfer money
- Social Engineering to generate a sense of urgency
- Easy + Cheap attack = Big potential reward

> *Relies on the oldest trick in the book*
> **Deception**

How it works — Create a fake account

How it works — Supply the CEO's Info

How it works — Deliver the attack

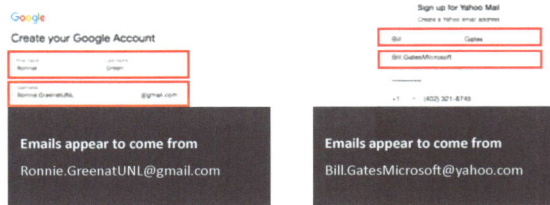

Emails appear to come from
Ronnie.GreenatUNL@gmail.com

Emails appear to come from
Bill.GatesMicrosoft@yahoo.com

Our Hacker

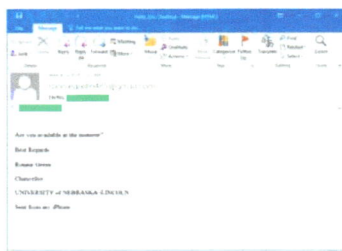

Email 1

- Sender address
- Am I available?
- Why is the sender using a non-university email account?
- Goal is to get a reply from you

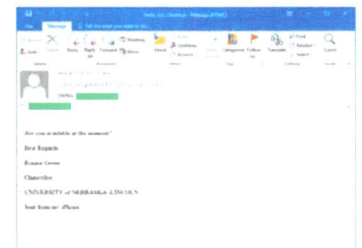

Email 1

- Sender address
- Am I available?
- Why is the sender using a non-university email account?
- Goal is to get a reply from you

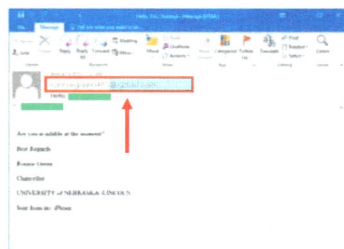

Email 2

- Sender details
- Subject line
- Why scratch off the back?
- Sense of urgency
- I want a $500 gift card

Email 2

- Sender details
- Subject line
- Why scratch off the back?
- Sense of urgency
- I want a $500 gift card

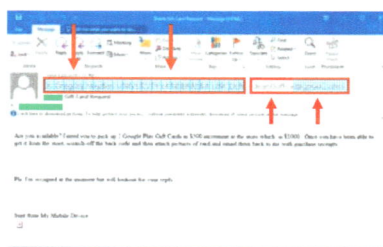

How to respond

- Don't 'play with them' – report them to security@nebraska.edu
- Pick up the phone and call the sender to verify.
- Any response indicates the trick was good enough to yield a reply

> Any response proves the attack was successful.
> Someone responded.
> Motivation to keep trying.

Pssst....YOU are the target

Email Statistics

- Data covering the past 30 days at NU *(all-campuses)*
  - 63,824,130 – Unique Emails
  - 49,419,576 (77%) -- Blocked as spam or malicious emails
  - 14,404,555 (22%) -- Continued to users as not spam

Phishing Protection

- Enterprise Protection
  - Sits in front of our Mail service and scans inbound and outbound traffic
  - Scores email based on content and reputation
- TAP (*Targeted Attack Protection*)
  - Protects against targeted attacks.
  - Scans URL's and Attachments
  - Alerts when things initially get past Enterprise Protection

TAP Dashboard

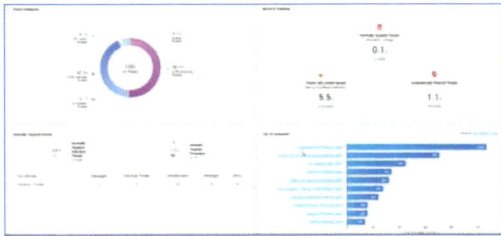

Upcoming Improvements

- Proofpoint TRAP(Threat response auto pull)
  - Not all malicious email gets stopped by proofpoint. Highly targeted attacks usually make it past at first.
  - TAP will detect but the malicious email still sits in the inbox
  - TRAP works with TAP and allows us to pull that malicious email back.
  - Limits the scope of the attack
  - Provides metrics and alerts on what made it through and to who.

TRAP

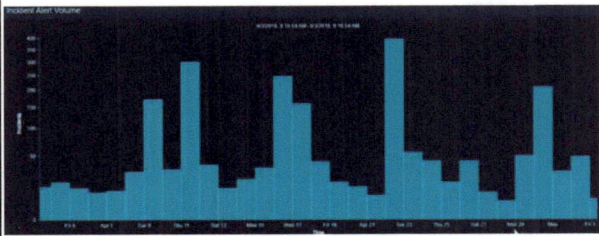

Upcoming Improvements

- Proofpoint CLEAR (Close Loop Email Analysis and Response)
  - Automates reporting and analysis of malicious emails.
  - Single solution for reporting suspected malicious emails to the security team and to proofpoint.
  - By clicking a report phish button in your mail client the email's headers and information is sent to the security team for analysis
  - It also sends a copy to proofpoint and trap for sandboxing and if enabled can allow for automatic pulling of that phish email out of inboxes.

Phish Button

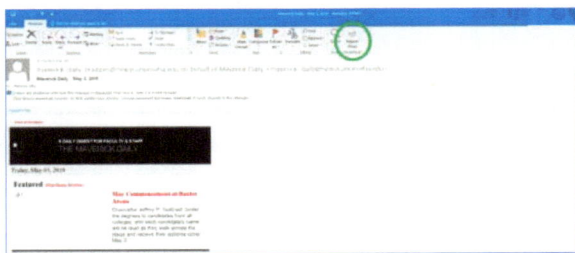

How to protect?

- 2 Factor (also known as multi-factor, MFA)
  - Something you have
  - Something you know
  - Something you are

https://its.unl.edu/services/duo/

Questions?

security@nebraska.edu

Featured Extended Presentation

Redesigning Courses & Determining Effectiveness Through Research

Tanya Joosten, University of Wisconsin-Milwaukee (UWM),
Erin Blankenship, Ph.D. (UNL),
Ella Burnham (UNL),
Nate Eidem, Ph.D. (UNK),
Marnie Imhoff (UNMC),
Linsey Donner (UNMC),
Ellie Miller (UNMC)

The second half of this extended session will build off of the information gathered in the first half of the session. During this session, audience members will break into small group discussions with members of the panel to discuss the re/designs of their course and research while comparing notes with the efforts undertaken by the panelists.

Jupyter Notebooks: An On Ramp for Advanced Computing & Data Science Resources

Carrie Brown (UNL)

David Swanson, Ph.D. (UNL)

Jupyter Notebooks are interactive documents that interweave live programming code with rich text elements making them ideal for educational and presentational applications. The inclusion of formatted text, equations, figures and links along with editable portions of code called "cells" allow the use of these notebooks for self-directed tutorials and demos. Course notes may be interweaved with dynamically computed coding exercises to lead students through pre-arranged demos and eventually to enable the learner to explore beyond the original exercises within the same course templates. Notebooks can be used as presentation slides, exported as static HTML or distributed in their native format so students can modify and create code independently or through directed exercises. The Holland Computing Center hosts easily accessible Jupyter Notebooks that will be presented and demoed during this presentation.

This presentation featured:
- Jupyter Notebooks and how they are used to improve online education.
- data analytics and high-performance computing.

Jupyter Notebooks
An On Ramp for Advanced Computing and Data Science Resources

Agenda

- HCC Overview
- User diversity at HCC
  - no longer familiar with CLI
  - Workshops to teach CLI
  - Alternate interface - Jupyter Hub

Pop Quiz!

1. How many here have heard of HCC?
2. How many here have ever logged into an HCC resource?
3. How many have used a CLI?
4. What did you come here to learn?

Thank you!

HCC cannot function without help from ITS — here are a few examples:

- WAN upgraded to 100 Gbps in a joint NSF proposal between ITS & HCC
- Our data centers are connected via fiber maintained by ITS
- HCC leverages Duo, provided by ITS
- HCC has equipment recently installed in WSEC29
- ITS & HCC have joint employees and serve on each other's hiring committees
- This doesn't happen everywhere and HCC is grateful!

> "HCC operates as a core facility and provides services that are critical to the research productivity of faculty and students across the University of Nebraska campuses."
>
> *–HCC 5-year Review, 2016*

> "HCC exists to reduce the time to science."
>
> *–David Swanson*

HCC is good at …

- **Big Data**: We manage over 10,000 TeraBytes (10PB)

- **Scaled Out Computing**: 35,000 x86 cores, 200,000 cores with GPUs, OSG, advanced networking, high bandwidth transfers, Globus Online

- **Scaled Up Computing**: Parallel processing, XSEDE, GPUs

HCC has great people

- Most are computational scientists

- 7 System Administrators

- 6 Application Specialists

- 2 Full time research personnel

- Favorite collaborative tools: table with a whiteboard

- Collaboration, not outsourcing outfit

- Significant outreach via Kickstarts and Software Carpentry

HCC has great people

- 118 Schorr Center

- 152 Peter Kiewit Institute

- Buffet Cancer Center 5.12.397

- hcc-support@unl.edu

- http://hcc.unl.edu

HCC Overview, Research Resources Board, UNMC, Jan 14 2019

HCC is free*

- *Shared usage of resources:* no charge

- *Best effort support:* no charge

- *Dedicated computing, storage or support:* up-front fees

  to many users! Paid for by NRI, NSF, NIH and NU researchers

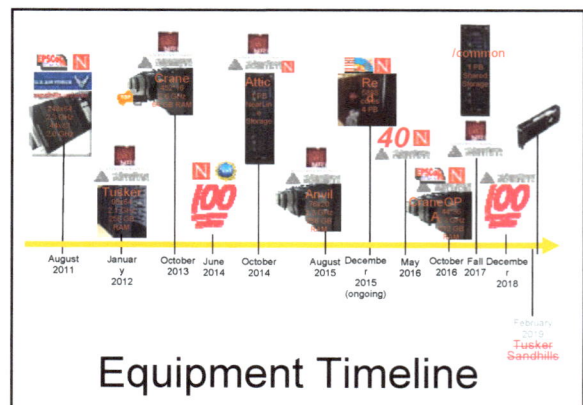

10 PB, 35,000 cores, 100 Gbps
1,045 active users, 237 research groups
OSG, XSEDE
Tier 2 site for U.S. CMS
iRODS, HDFS, Parallel Filesystems

Equipment Timeline

Anvil

- Customizable virtual machines
- For projects not well served by a traditional Linux environment:
  - interactive environments or alternate operating systems
  - projects that require root access or dedicated resources
  - test cluster environments

Attic

- Near-line data archive
- Backed up in Lincoln and Omaha for disaster tolerance
- 10 Gb/s transfer speed to and from the clusters when using Globus Connect
- Cost lower than commercial cloud services

Thank you ITS!

High performance network transfers
81.11 Gbps current speed record

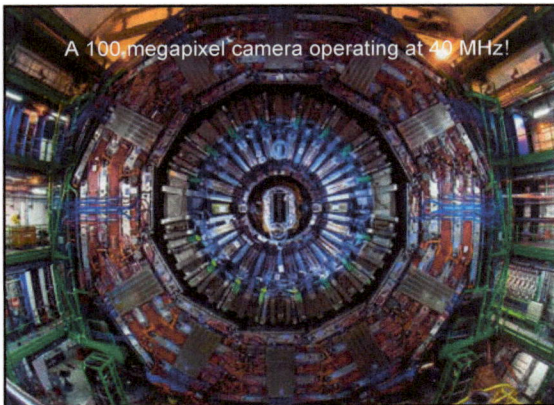

A 100 megapixel camera operating at 40 MHz!

Science is a Team Sport

Holland Computing Center aids gravitational wave discovery

User Diversity at HCC

As of October 2018:

- 77 departments system-wide

- 237 research groups

- 1045 users

Usage Last Year

	Department	CPU Hours: Total
1	[UNL] Chemistry	32,350,065
2	[UNL] Mechanical and Materials Engineering	19,091,323
3	[UNL] Arts and Sciences	16,295,027
4	[UNL] Physics and Astronomy	16,068,962
5	[UNL] Chemical and Biomolecular Engineering	14,247,323
6	[UNL] Center for Biotechnology	11,136,059
7	[UNL] Computer Science and Engineering	3,974,968
8	[UNMC] Pharmaceutical Science	3,065,034
9	[UNO] Physics	1,781,838
10	[UNL] Educational Psychology	1,695,276

Usage Last Year

Department	CPU Hours: Total
11 [UNO] Economics	1,452,076
12 [UNL] Earth and Atmospheric Sciences	1,365,949
13 [UNL] School of Biological Sciences	1,327,096
14 [IANR] Agronomy and Horticulture	991,450
15 [IANR] School of Natural Resources	883,332
16 [Creighton] Genetics Cell Biology and Anatomy	797,908
17 [IANR] Animal Science	697,021
18 [IANR] Statistics	515,787
19 [UNL] Finance	321,137
20 [IANR] Survey Division - School of Nat Res	300,916

*77 Departments Total

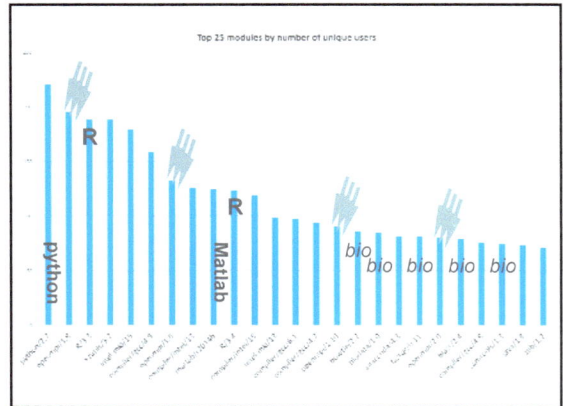
Top 25 modules by number of unique users

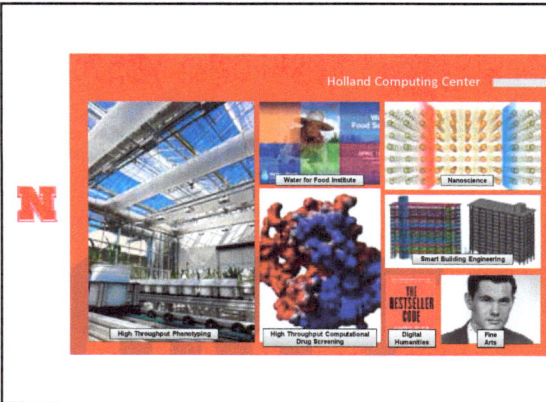
Holland Computing Center

Software Carpentry at HCC

- "Teach skills needed to get more done in less time with less pain"
- Community of volunteer instructors, maintainers and helpers
- 4 certified instructors
- Two-day "boot camp" style workshop using live-coding and hands-on exercises
- Three core topics: Bash, Git and Intro to Programming (R/Python)

Other HCC Training

- Open Office Hours
- Other workshops: QIIME2, Fall Kickstart, June Workshop Series
- Course support: Deep Learning, Statistics, others
- Group + whiteboard
- Rounds
- Semester courses: Cluster & Grid Computing, Parallel Programming

Attendees

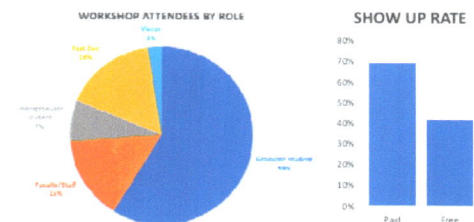
WORKSHOP ATTENDEES BY ROLE   SHOW UP RATE

Workshop Effectiveness

- Tracking cluster utilization over account lifetime (measurement of productivity!)
- Workshop attendees and matched pair
  - Account creation date, Position, Department
- Preliminary results promising
  - Ave. Lifetime CPU Hours Used 49,159 vs 1,549,614 (p=0.0544)

CLI

- HCC runs primarily LINUX clusters
- Traditional interface is a BASH Shell
- Users interact via a text-based prompt
- submit jobs via a script (we use SLURM)
- results written to a file that users must know how to read and/or edit
- Limited "real-time" functionality

CLI Interface

Jupyter Hub

- Embed formatted text with live code boxes
- large community (SWC included)
- Web enabled
- Can be used interactively or asynchronously
- Popular kernels and support for custom libraries

Jupiter Hub Demo

Acknowledgements

- ITS
- University of Nebraska
- Nebraska Research Initiative
- Holland Computing Center
- NSF, EPSCoR, DOE, NIH

Using Backward Design & Authentic Learning to Build Curricula from Competencies

Christine M. Arcari, Ph.D. (UNMC)

Analisa McMillan (UNMC)

UNMC College of Public Health was faced with big changes in accreditation criteria and found the need to move from faculty-centered teaching and discipline-based content towards student-centered learning and outcome-based curriculum that was integrated and built around competencies. UNMC began by mapping competencies to courses, but a big question kept resurfacing was "how do we assess the competencies we're mapping?" Learn how UNMC used backward design to create new courses and existing courses through authentic assessments to meet the new competencies.

This presentation featured:

- Backward Design Process and how it relates to course design.
- examples of how to gain faculty buy-in to this new format
- authentic learning assessments.

1

(CMA) Thank you Mollie for introducing us and inviting us to share the work that we have been doing to build our curriculum and design our courses based on competencies.

I'm Christine Arcari

(AM) And I'm Analisa McMillan

(CMA) Today we will be presenting the backward design process that we are using here in Nebraska to redesign our curriculum and courses.

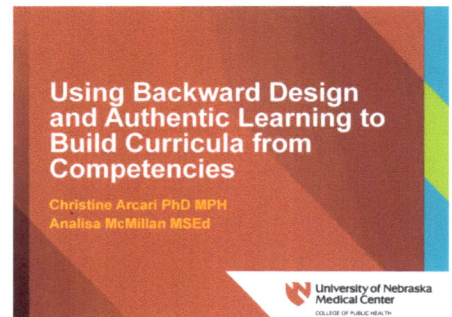

2

(CMA)

I've put this slide here as a reminder for where we started back in November 2016 when the revised CEPH accreditation criteria were released. There were so many questions...

From the big picture - what is this new framework for assessing the quality of public health education? Down to the microscopic level starting with - what is a competency?

There is no single answer to these questions – we all face a multitude of choices and we all are figuring out the best path for our schools and programs.

3

We began navigating our way by mapping competencies to courses. But a big question we kept coming back to was how do we assess these competencies we're mapping. Our big idea was to use the Backward design process and authentic learning as we designed new courses and refreshed our existing courses.

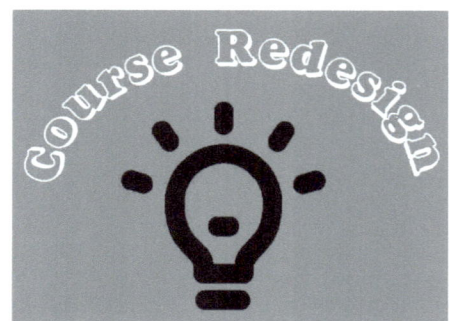

4 (AM)

Backward design compels us to design and create courses based on competencies and not on the books, content or activities that we were most comfortable with. Competencies are the framework that guides our design and identify both teaching and learning experiences. Backward design is different from some of the more traditional models that start out with writing learning objectives, then moves on to the planning of course content and activities, and finishes with figuring out how you will assess the students. As you can see the traditional methods often focus on the teaching and not on the learning aspect of the course. Backward design helps you focus on the learning aspect of the course - what the learner needs to know and do to accomplish the goals.

When using backward design you begin by thinking about the end result, or in our case the C-E-P-H foundational competencies and our concentration competencies that students must meet to graduate from our programs. After you know where you are going, you will develop a plan to assess the competencies before moving on to the delivery of the knowledge and skills that we provide as teachers.

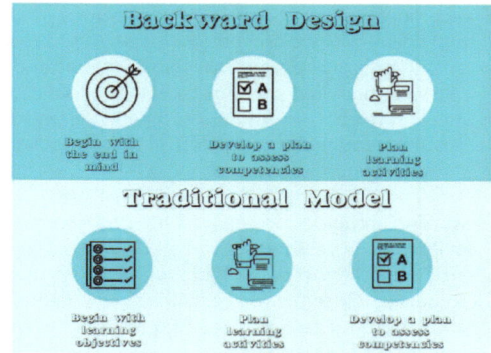

5 (AM)

The three steps to backward design are to first identify the desired results. This is the big picture, what do you want your students to know and be able to do after they complete your program? The second step is to determine the assessment evidence. How will you know what students know and can do once they graduate? The last step is to plan the learning experiences so that you can make sure that students are being taught the knowledge and skills to meet the assessment criteria.

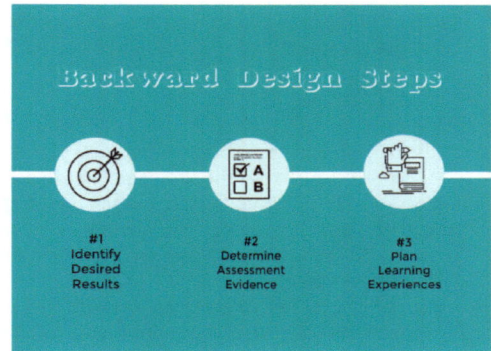

6 (CMA)

Now that we have an overview of the process, let's discuss Step 1. Identify the desired results. This is the step where you ask what should the students know, understand and be able to do at the end of the program. In this instance, the desired results are known –these are our competencies. CEPH provides foundational competencies for the MPH and DrPH degrees. We provide additional competencies for each MPH and DrPH concentration. And we also must provide competencies for MS and PhD concentrations. In this webinar, we are going to focus on the MPH foundational competencies. We used this same process to design courses for all of our degrees and concentrations.

To design our MPH core, the desired results were the 22 CEPH foundational MPH competencies.

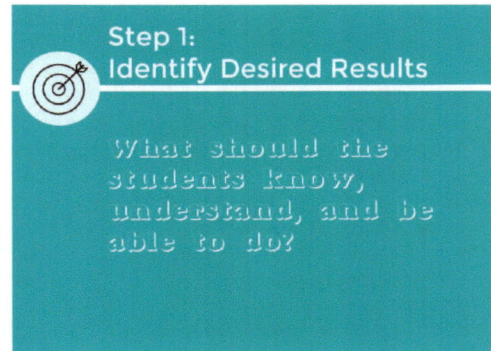

7 (CMA)

We will discuss briefly the process we used to build our MPH core courses and we will go more in-depth about the process of designing a course to assess competencies.

Here is a look at our process to build the MPH core....

We appointed an MPH redesign workgroup made up of faculty representing all 5 departments in our college, and Brandon Grimm, our Director of Masters Programs, Analisa, our Instructional Designer and myself the associate dean of academic and student affairs. I locked everyone in a room with enough food and water to last 3 days. I am totally just kidding, I only threatened to do that and didn't have to because we have a great team here. What we did is commandeer a large classroom with lots of wall space. We printed out all of the foundational competencies separately and realized they were too hard to read and process for this exercise so we printed the main theme for each competency and included the CEPH number in the bottom right corner. Everyone had a sheet in front of them with the complete wording for all competencies. Having this sheet to refer to was important as we found when we continually referred to the competency using the abridged description and did not periodically read the full text of the competency we experienced some drift in interpretation . The task given to the workgroup was – here are the 22 competencies that make up our core, clearly we will not be offering 22 core courses – how do we combine these 22 competencies to create the MPH core. As expected, this led to lots of vigorous discussion.

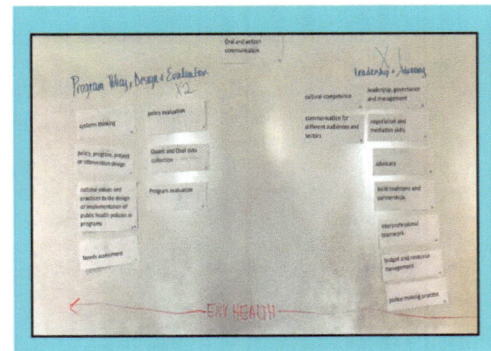

8 (CMA)

The competencies were grouped and regrouped and regrouped again and again and again until we arrived at a consensus. We started with a 2 hour meeting of the large group, which was both chaotic and productive. Importantly, faculty were prepared and familiar with the foundational competencies before the meeting began. After that first meeting, Brandon Grimm, our Director of Masters Programs and myself checked in with each other to reassure ourselves that the large group was considering all the issues that needed to be considered. It took one more large group meeting that lasted 2-3 hours reach a consensus. The working group decided on a 15-hour core comprised of 5 courses – Foundations of Public Health, Biostatistics, Epidemiology, Planning and Evaluation, and Leadership and Advocacy. All 22 foundational competencies were mapped to these 5 courses. Once the 5 core courses were identified, a backward design process and a focus on authentic learning was used to design all the courses. This process worked well for courses in various stages of development. The biostatistics and epidemiology courses were well-established courses that needed to be refreshed; the Foundations of Public Health course was also an existing course, but needed a major redesign; and the Planning and Evaluation Course and Leadership and Advocacy course were new courses that needed to be developed. We found the backward design process was helpful for all these scenarios.

9 (CMA)

Once the 5 core courses were agreed upon, a multidisciplinary workgroup was created for each course. At the first meeting, each workgroup was tasked with letting go - and told NOT to think about the design of the course based on the textbook that would be used, or the way they were taught the course as a student (probably 20 years ago), or – and this was the hardest one – NOT to use the current syllabus if there was one.

How did we get to Yes? We got to yes with lots of faculty development workshops, creating buy-in, and continuous communication with faculty throughout the process.

10 (CMA)

So we want to share some tips with you that we learned in step 1.

#1: There is no single correct way to design the MPH core courses. There are so many permutations on how to group these competencies and I think we went through about half of them. Just because a grouping looked good taped to the wall, it didn't mean it was the right grouping of competencies for our school and our students. Some problems that we encountered: Thinking too big or too small – for example a course with 18 competencies and a course with 1 competency. Designing a course that no faculty members wanted to teach – Ok here's a great combo of competencies, now who's going to teach this course? and suddenly no one is looking you in the eye. It's important to maximize the resources that you have available to you. Also, it's important to not lose sight of your students – who they are and what you are training them to do.

#2: Faculty training is essential. We worked to educate all faculty in the college, not just the faculty participating on a working group. We covered such topics as: What is the difference between a competency and learning objective? What is backward design? What is authentic learning? What is a rubric? Why are we doing this? And the popular – Introducing the New Couse Syllabus Template.

#3: CEPH is a great resource. Even though we has just finished a site visit under the old criteria I attended the accreditation workshop held by CEPH and it was super helpful particularly in understanding the intention and interpretation of the competencies and reporting. Having a firm understanding was helpful during group faculty discussions which inevitably would get hung up on semantics such as "Does and mean and or does and mean and/or". And for the record – and means and NOT or. If in doubt – contact C-E-P-H.

and finally, #4: Understand that most if not all of your faculty were not trained in a program driven by core competencies. Change is hard. And it's imperative to have a facilitator to redirect and make sure the group doesn't get lost down a rabbit hole. For example, I was seriously concerned that banning the current syllabus from the course redesign groups was going to lead to serious acute and chronic health conditions among group members. Sometimes, we just had to let the group talk about the current syllabus, get it out of their system and then cut off the discussion by saying great discussion – now let's forget it and move on.

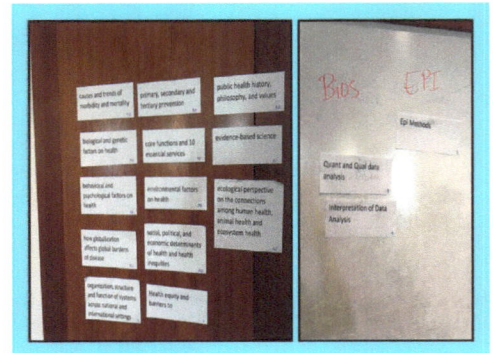

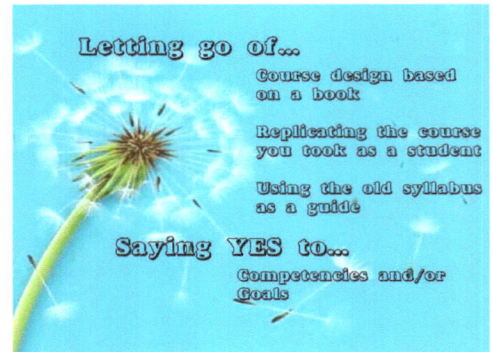

Letting go of...
Course design based on a book
Replicating the course you took as a student
Using the old syllabus as a guide
Saying YES to...
Competencies and/or Goals

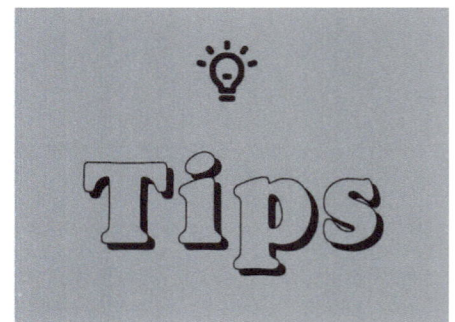

Tips

11 (AM)

The next step is to determine the assessment evidence. How will we know if students achieved the desired results? What will we accept as evidence of student understanding and proficiency? In this step we start to think of a course in terms of evidence we collect from assessments and how we can document and prove that the desired learning outcomes have been achieved. I like to think of this step as the bridge between the course goals and your teaching. When designing our curriculum this step became the most crucial step because it not only showed us how we would demonstrate that our students were meeting a competency, it informed our teaching and content delivery for each course.

12 (AM)

When designing our course assessments, we wanted to infuse each course with using authentic learning assessments, that could be assessed on an individual level (even when it is a group project), to demonstrate competency attainment. We want our students to experience the "real world" with hands on, real world scenarios that test that our students have not only acquired the knowledge needed but can also apply the knowledge. We want our students to be practice-ready when they graduate. During the very first course workgroup meeting, we discussed authentic learning and the 10 design elements and why we should include authentic assessments in each course. As you can see on the slide the elements are to design assessments that use real-world relevance and Ill-defined problems. Create assessments that contain sustained investigation and multiple sources and perspectives. Build collaboration, reflection, and interdisciplinary perspective into assessments. Create integrated assessments, polished product assessments and assessments with multiple interpretation and outcomes.

13 (AM)

After discussing the ten elements, we tasked the workgroups with creating authentic assessments by thinking outside the traditional methods that are often seen in courses. We let them know that traditional methods were not off the table and that for this session we would be creating authentic assessment ideas that may or may not be included in the final course. To aid their creativity, they were supplied with a list of examples of authentic learning to include things like multi-media creation, field trips, design projects, concept mapping, case studies, timelines, real data sets, infographics, peer editing and more. During our course workgroups, faculty referred to this list when brainstorming and shared other ideas of authentic assessments with their peers. Many of the workgroup members were excited to know that they were already including authentic assessment in their other courses and were excited to add more.

14 (AM)

During the first-course workgroup meetings, each workgroup was asked to brainstorm assessments for each competency mapped to the course. To make it easier on the faculty, we placed the competencies on the wall as you can see in the picture, and we worked our way through them one by one sharing ideas for authentic assessments. We wrote each idea on the papers on the walls, to keep everyone focused and everything in one place. There were 2 guiding principles to this exercise – generate as many ideas as possible for each competency and be creative. We asked them not to think about the current course or the current course assessments during this brainstorming session but we did tell them that they could add them to the list after we generated some new ideas. Surprisingly the new assessment ideas were so good that they did not rely on the old traditional assignments. We usually had more than one assessment idea per competency and selected the one or two assessments that fit best for the course. Sometimes, we combined multiple assessment ideas that fit well together into one assessment or we created an assessment to fit more than one competency. This process took one to five hours depending on the group dynamics and the number of competencies aligned to the course.

15 (AM)

After the workgroup SELECTED the assessments TO MEASURE COMPETENCY ATTAINMENT, we spent time developing the course learning objectives. The goals of the learning objectives are that they should be specific measurable statements that are written in behavioral terms and let us know if an assessment criteria is met. Once we aligned the completed learning objective with the competency, we would know the competency was met. The assessments and learning objectives are also used inform the teaching and content delivery of the material. When we created learning objectives for our MPH core courses, we aimed for the apply, analyze, evaluate and create domain levels targeting more complex and abstract learning.

Step 2: Determine Assessment Evidence

How will we determine if the desired results occurred?

Ten Design Elements of *Authentic Learning*

Examples of Authentic Learning

Authentic Assessment Ideas

Verbs matter... Bloom's Taxonomy

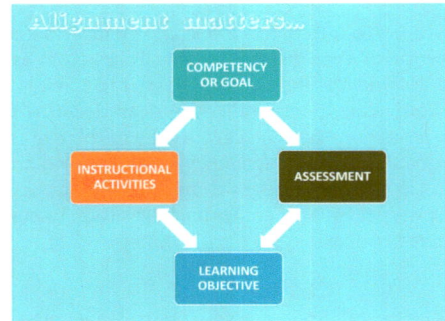

17 (AM)

This table is an example of our course competency mapping. As you can see the assessments are mapped to the learning objectives and competencies. By creating this table we can see if we are covering all of the competencies that are mapped to the course and determine if we have too much or too little coverage on any one competency and if we needed to refresh, rearrange or redesign our assessments.

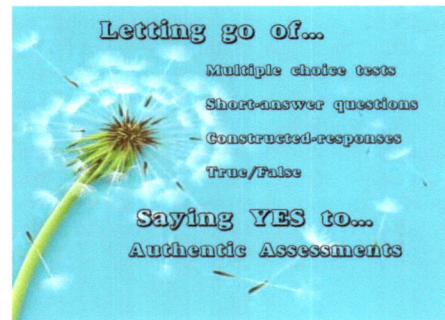

18 (CMA)

Once the 5 core courses were agreed upon, a multidisciplinary workgroup was created for each course. At the first meeting, each workgroup was tasked with letting go - and told NOT to think about the design of the course based on the textbook that would be used, or the way they were taught the course as a student (probably 20 years ago), or – and this was the hardest one – NOT to use the current syllabus if there was one.

How did we get to Yes? We got to yes with lots of faculty development workshops, creating buy-in, and continuous communication with faculty throughout the process.

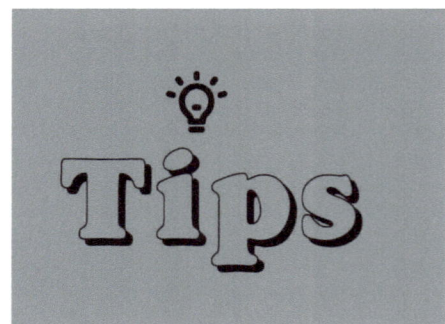

19 (AM)

Some additional tips for this step are to remember that Backward design requires a shift in thinking. Sometimes you have to redirect faculty back to the task. It may also be a challenge for some faculty to focus first on the desired results and let the teaching follow.

Another tip is to create learning experiences for faculty that "chunk" or break down the steps of the backward design process. As mentioned earlier we conducted sessions on writing objectives, using the new syllabus, competency mapping and creating rubrics. It is a big task, so make it manageable.

And my last tip is during the first workgroup take time to explain the backward design process and the reason behind it. You may have faculty ask a lot of questions and it is good to prepared.

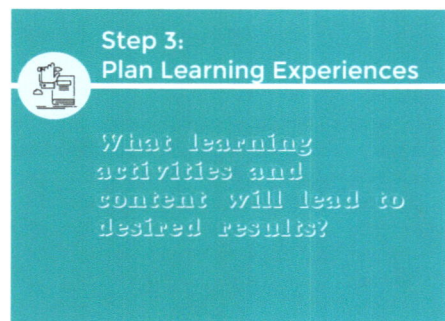

20 (AM)

The last step is to plan the learning experiences and instruction. What learning activities and content will lead to desired results? We have identified the results and the appropriate evidence. Now it is time to think about the most appropriate instructional activities and content delivery. There are some key questions that need to be considered and they are. What facts, concepts, principles, processes, procedures, and strategies will students need to achieve results and perform effectively? What authentic activities will equip students with the needed knowledge and skills? What will need to be taught and how should it be taught? And what materials and resources are best suited to fit these goals? We want to clearly define what activities, lectures, and content will be designed to meet the learning experiences necessary to help our students be successful in meeting the competencies.

21 (AM)

Remember all the discussion we had about authentic learning assessments? Now is the time to pull out that list to see if any of the ideas can be adapted for the classroom. Design individual and group activities that take place in and out of the classroom. Identify existing case studies or create your own real-world case studies that allow students to apply their knowledge. Locate real world data and examples to use or charge the student with finding data and using it just as they would have to if they were in the "real-world". Create discussion prompts to be used in the classroom or online setting that promote meaningful discourse amongst students and the faculty. This is also the time to identify a textbook and/or individual readings.

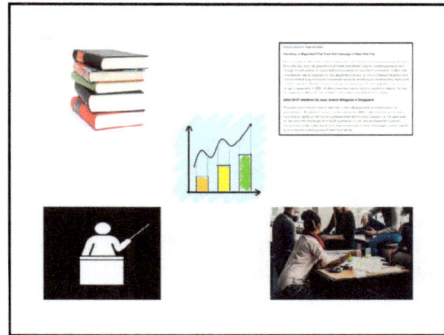

22 (CMA)

Now I want to dig a little deeper into course assignments and grading. I'm going to use our Foundations in Public Health course as an example. Our Foundations course maps to the 10 foundational learning objectives and 2 MPH foundational competencies.

I'm going to include in this example building assessments from a foundational competency and a foundational learning objective. However, I want to be clear that C-E-P-H requires that we report specific assessment opportunities only for competencies and not for the foundational learning objectives. We decided, for our college, that it worked well for us to build specific assessments for the foundational learning objectives.

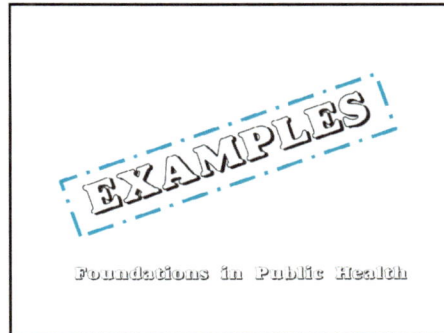

23 (CMA)

First, in every course syllabus – there is a map that lists the competencies addressed in the course, the learning objectives in the course mapped to these competencies, and the assessment product. For simplicity – I'm only showing you only 1 competency – but there are 2 competencies mapped to this course. In fact, I've simplified the next several slides to enhance understanding.

As you can see down the first column – we created a master ID for each competency for ease of tracking. Here we have MPHF6 – this refers to the MPH CEPH Foundational Competency #6. We did this because competency #1 in one syllabus could be the same as competency #3 in a different syllabus and that gets confusing.

24 (CMA)

Each assignment included in the competency mapping table is described in more detail in the syllabus under course assignments.

For example, MPHF6, instead of an exam question asking students to define structural bias, social inequities and racism, the assignment has students choose and complete three of the Harvard implicit bias assessments. Students must then write a reflection paper answering several questions as shown on the slide.

Remember it may take more than 1 assignment to demonstrate competency attainment. For example – this competency could be met using 3 different assignments – each one focusing on just one of the following: structural bias, social inequities, or racism. Or there could be multiple assignments that focus on all those concepts collectively but differentiate organizational, community and societal levels.

25 (CMA)

Here is another example, but this time FLO7 – which refers to foundational learning objective 7. Again, we do NOT have to report how we are assessing the foundational learning objectives – we just go with the FLO and list the course or courses where the FLO is covered. We use the backward design process on all elements of the course. Usually, the desired results we identified were the competencies. However, there were instances when we also wanted to build an assessment around a course learning objective which does map to a competency. The foundational learning objectives are a good example of this.

26 (CMA)

Here are three examples of water or air 5 picture stories explaining the effects of environmental factors on population health.

27 (CMA)

Let's take a closer look at one of the examples – Flint Michigan Water Crisis and Learning Issues:

This 5 picture story shows the story of the Flint water crisis. In April 2014, the city of Flint, Michigan changed its water source from the Detroit Water and Sewerage Department to the Flint River. The water from the Flint river was not chemically treated with corrosion control measures. This was problematic because the water entered houses via lead pipes, and lead leached from the pipes in to the water supply. High exposure can lead to lead toxicity both acutely and over time. This is particularly concerning because of the negative health effects of lead exposure on children. The public health response to the Flint MIchigan Water Crisis included supplying clean water to residents in the city and testing children for lead exposure.

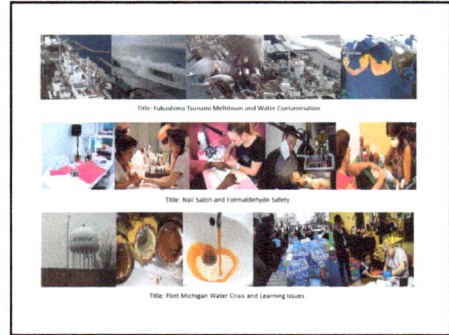

28 (CMA)

And here is the rubric.

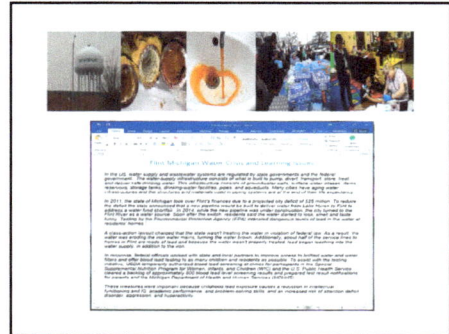

(1 – 2 – 3)

29 (CMA)

Once the 5 core courses were agreed upon, a multidisciplinary workgroup was created for each course. At the first meeting, each workgroup was tasked with letting go - and told NOT to think about the design of the course based on the textbook that would be used, or the way they were taught the course as a student (probably 20 years ago), or – and this was the hardest one – NOT to use the current syllabus if there was one.

How did we get to Yes? We got to yes with lots of faculty development workshops, creating buy-in, and continuous communication with faculty throughout the process.

30 (CMA)

And finally our tips for step 3:

#1: Keep the list of authentic assessments that you created in step 2 and use them to create activities in your class that allow the students to apply knowledge and not just receive information. You created authentic learning assessments so why not continue that in the classroom by being open to new ideas and authentic learning experiences.

#2: Using multidisciplinary teams to design the core courses increased creativity and best practices across disciplines were shared.

#3: Putting thought in to tracking for reporting purposes and coming up with a plan to collect the information needed for reporting is essential - and the reason why we designed a comprehensive course syllabus template.

#4: Some faculty are not familiar with grading rubrics. And it's hard to put a rubric together for an assignment that you have no experience using previously. Course directors have been told to expect that if the grading rubric included in the syllabus is untested for the assignment, it will most likely need to be revised a bit – and course directors will be encouraged to make updates for the following year as the semester progresses instead of waiting 6 months before you start planning for the next time you teach the course and can't remember anymore what worked and what didn't work.

And #5: A very important lesson is to make sure the working groups include the faculty that will be teaching the course. If more than one faculty member is teaching the course throughout the year, these faculty members need to come to a consensus. Faculty members may differ in their teaching style and examples used –some course directors might even reorder the presentation of topics. But it is imperative when multiple faculty are teaching the same course throughout the year, that the competencies mapped to the course are being covered and assessed.

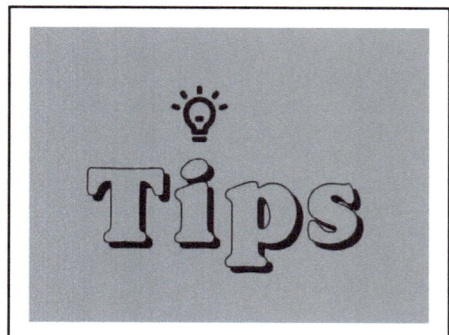

31 (CMA)

So we consider our backward design process a success at our college. We provided the big picture to the faculty before drilling down in to the details and course design to provide context because we know change is hard.

Once it is determined which competencies are mapped to the course, it took about 4-8 hours of workgroup time to identify authentic learning assessments for each competency and 5-20 additional hours to develop the course syllabus.

It was very important that this was a collaborative process across departments. As you may have noticed there is not a dedicated Environmental and Occupational Health course in our core. But we made sure environmental health had faculty representation on all the working groups --- and all the core courses from biostatics to leadership and advocacy are infused with environmental health content.

Overall, the faculty, although at times frustrated by the process – liked it. It brought faculty together to share ideas and experiences and it was an opportunity to work with colleagues that you might not ordinarily interact with.

My own personal belief and experience is that all the extra work put in upfront in developing a course really pays off while teaching the course. It's like no longer trying to build the plane while flying it and that's a good feeling. And having a grading rubric reduces the time it takes to grade an assignment and promotes consistency in grading.

And a final key to success was the review process. Once the syllabi were complete for the 5 core courses, comments were sought from all key stakeholders - faculty, staff, and a portion of our students, alumni and employers. All comments were brought back to the working group and carefully considered and changes were made before sending the course to be reviewed and approved by the curriculum committee. This review also had the added bonus of generating lots of enthusiasm for the revised curriculum giving us good momentum as we move towards the Fall 18 kick-off.

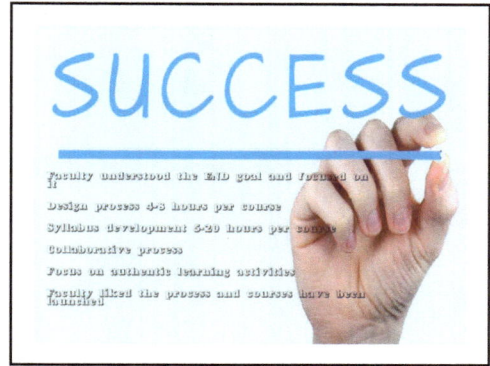

32 (AM)

We would like to take a moment to recognize our faculty and staff who helped enormously in this process. A big thank you to Brandon Grimm who co-chaired the committee, the MPH core redesign group and the 5 core course design/ revision groups. Without their involvement this process would not have been possible.

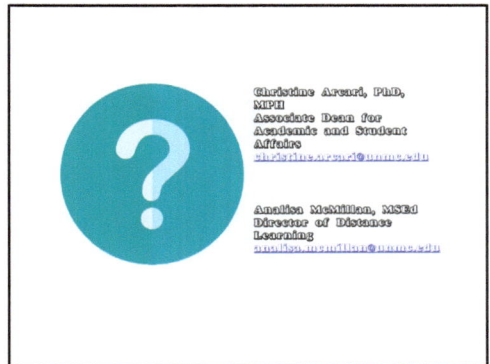

33

Now we are happy to answer any questions you have.

34

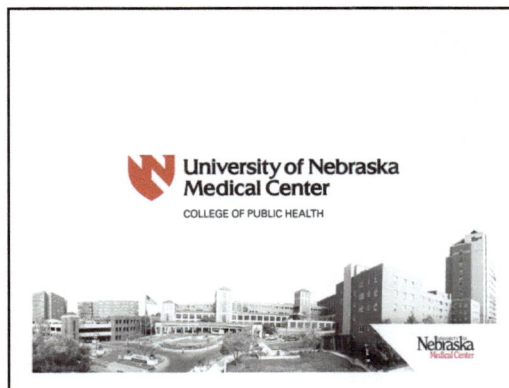

Creating, Building & Nurturing an Online Program: A Success Story

Melissa Cast-Brede, Ph.D. (UNO)

Jaci Lindburg, Ph.D. (UNCA)

Erica Rose (UNO)

Alex Zatizabal-Boryca (UNCA)

This presentation will present a case study of UNO Library Science program's migration to a fully online format. Shared information will include a step by step review of the creative process, the curriculum conversion process, outreach and retention strategies and plans for sustained growth. Discussions will cover collaborations with key partners, marketing plans, faculty structure, technology tools and a few words about attitude. In addition to meaningful dialogue, we will share resources (templates, data charts and presentations) in the hopes that others might be inspired about the potential and power of online learning. Presenters will answer any and all questions about why we made the leap to a fully online program, what steps we took to make it happen and how we balance growth and quality instruction.

This presentation featured:
- the planning process for designing a fully online program.
- explanation on how intentional lead nurturing can affect retention.
- tools and procedures used in designing and implementing online curriculum.

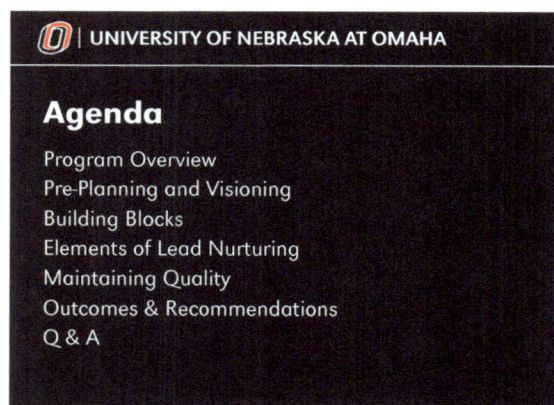

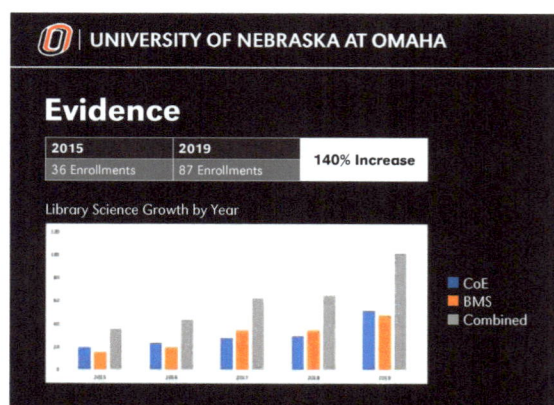

Slide 1

Evidence

- 48% of LibSci Students are out of state
  - Georgia, Florida, Nevada, Kansas, Texas, Iowa, Maryland, South Dakota, New York, Idaho, Illinois, Michigan, Minnesota, Colorado, North Carolina, Colorado, Arizona, California, Wyoming, Virginia, New Jersey, Idaho, New Hampshire, Nebraska, Delaware, Bermuda, Canada

Slide 2

VISION

Slide 3

Decision factors

- Accessibility to quality education
- Niche market
- Market research
- Program needs/professional needs
- Trajectory of higher education

Slide 4

Preparing for Online Delivery

- Office of Digital Learning
- Instructor Readiness
- Course Availability/Rotation Plan
- Integration
- Ongoing Resources

Slide 5

Planning

- Partners
  - Teacher Education Leadership
  - Advisors
  - Office of Digital Learning
  - NU Online
- Instructional strategy
- Program "flavor"
- Assembling the team

Slide 6

Purposeful Action

Spring 2016	Fall 2016	Fall 2017	2018-2019
Began online migration	Launched marketing campaign	Course migration complete	Nurturing a growing student body

Slide 7

BUILD

Slide 8

Quality Instruction

- Highly structured
- Consistent instruction (limited faculty to start)
- Slow conversion (Spring 2015-Fall 2017)
- Synchronous meetings
- Regular communication
- Project based learning
  - Developing critical thinking, creativity, research skills

UNIVERSITY OF NEBRASKA AT OMAHA

Standards

- American Library Association (ALA) Master's Accreditation
- American Association of School Librarians (AASL) Certified
- Interstate Teacher Assessment and Support Consortium (InTASC)
- Council for the Accreditation of Educator Preparation (CAEP)

UNIVERSITY OF NEBRASKA AT OMAHA

Curriculum Conversion

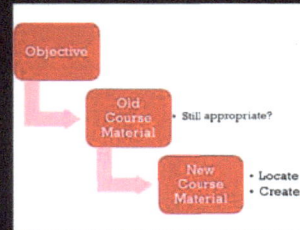

UNIVERSITY OF NEBRASKA AT OMAHA

Curriculum Conversion

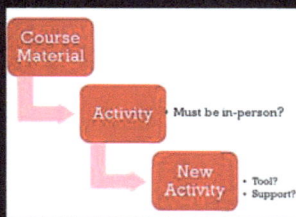

UNIVERSITY OF NEBRASKA AT OMAHA

Curriculum Conversion

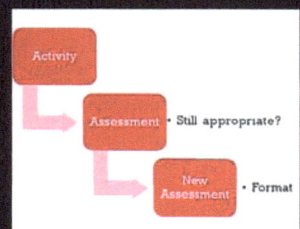

UNIVERSITY OF NEBRASKA AT OMAHA

Tools

- Learning Management Systems
- Collaborative Tools
- Video Creation Software
- Office of Digital Learning
  - Best practices and support for online learning

UNIVERSITY OF NEBRASKA AT OMAHA

Marketing

- Word of mouth
- Grassroots campaign

UNIVERSITY OF NEBRASKA AT OMAHA

NU Online Partnership

- NU Online portal
  - 36% of current students
- Social media/print advertising
- Conference booths
- National webinar

UNIVERSITY OF NEBRASKA AT OMAHA

NU Online Portal & Campus Integration

- Scope
- Tracking
- CRM Systems
- Advising Systems

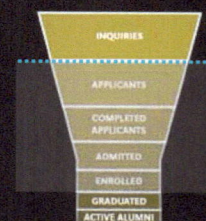

Slide 1

UNIVERSITY OF NEBRASKA AT OMAHA

NURTURE

Slide 2

UNIVERSITY OF NEBRASKA AT OMAHA

Lead Nurturing

- Quick Response Time – 24 hours
- Define Role and Scope
- Proactive non-mass outreach to encourage two-way dialogue
- Automatic Tracking, Follow up, and Reporting

Slide 3

UNIVERSITY OF NEBRASKA AT OMAHA

Setting Expectations

- Average time from Inquiry to Enroll is 6 – 8 months (undergraduate)
- 1 – 3 percent increase of historical performance (conversion rate)
- Multi-year conversion rates may be more realistic for certain programs

Based on Inside Track recommendations in Key Performance Indicators and Best Practices for University of Nebraska

Slide 4

UNIVERSITY OF NEBRASKA AT OMAHA

MAINTAINING QUALITY

Slide 5

UNIVERSITY OF NEBRASKA AT OMAHA

Instructional Quality

- Intentional hiring for adjunct faculty
- Adding sections and restructuring course calendar

Slide 6

UNIVERSITY OF NEBRASKA AT OMAHA

Quality, Cohesive Student Experience

- Team Approach
- Program and Academic Advising
  - Onboarding process
  - Program welcome wagon
  - Touchpoints throughout
  - Enrollment monitoring

Slide 7

UNIVERSITY OF NEBRASKA AT OMAHA

Quality in Evolving Technology

- Early Adopters
- Continuous Training
- OER
- External Resources/Partners

Slide 8

UNIVERSITY OF NEBRASKA AT OMAHA

Celebrations

- Program Numbers = 140% growth 2015-2019
- Online instruction (student satisfaction)
- High graduation and employment rates
- National visibility
  - # 1 in Best College Reviews

Recommendations

Vision Stage
1. Market Research
2. Identify Partners

Building Stage
1. Plan Conversion
2. Emphasize Consistency

Nurturing Stage
1. 24-Hour Turnaround
2. Define Role

Maintaining Quality
1. Prioritize Quality Instruction & Advising

Questions?

Erica Rose ecrose@unomaha.edu
Dr. Cast-Brede mcast@unomaha.edu
Alex Zatizabal Boryca azatizabal@nebraska.edu
Dr. Jaci Lindburg jlindburg@nebraska.edu

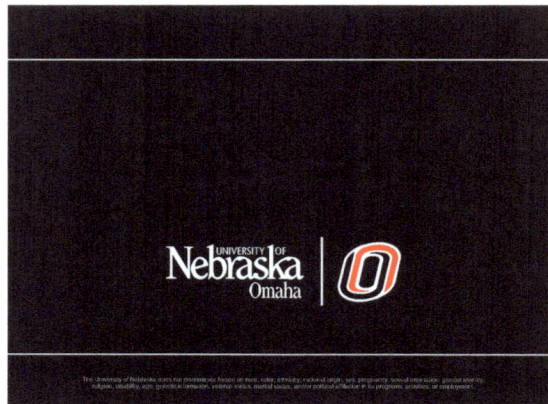

CIO Panel - Campus Updates

Mark Askren (UNCA)
Bret Blackman (UNCA)
Brian Lancaster (UNMC)
Deborah Schroeder (UNCA)

The CIOs will provide updates on ITS projects, campus issues and ITS successes. Bring your questions for Q&A time.

This presentation featured:
- updates on key ITS projects.
- updates on campus issues.

Educating with Technology Across Intergenerational & Intercultural Groups

Ogbonnaya Akpa, Ph.D. (UNL)
Toni Hill, Ph.D. (UNK)
Olimpia Leite-Trambly (UNK)
Sharon Obasi, Ph.D. (UNK)

According to Pew Research Center, there is a clear distinction between technology use and adaption across generations and racial and ethnic groups. The Pew research shows younger, white individuals possess and adopt technology more rapidly than older, non-white individuals. In contrast, there is limited research on the interplay between generational and cultural differences of university and college students and their learning with technology. Many modern educational strategies including interactive videos, game-based learning and multi-device learning address the ecosystems of resources used in combination with the methods of delivery, and the learning outcomes with less consideration given to the learner's background including generational age and cultural perspective. This presentation will look at both intergenerational and intercultural use of technology in educational settings and identify some promising strategies to reach and teach across diversity.

This presentation featured:

- intergenerational differences in use and adaption of technology.
- cultural differences in the use and adaptation of technology.
- new strategies to reach and teach diverse students.

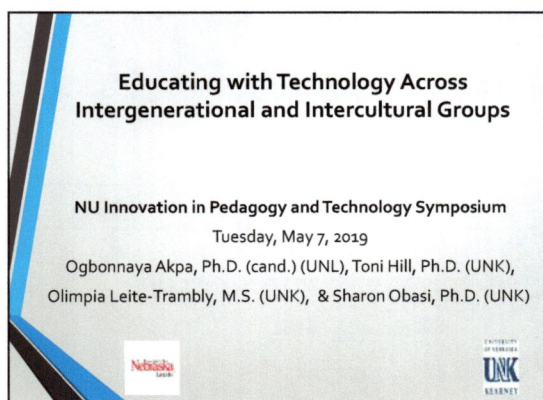

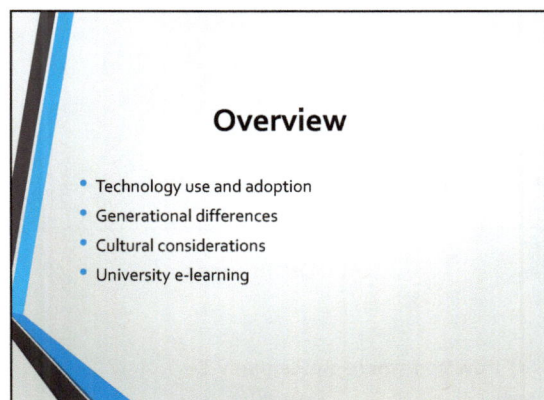

So, who are our students?

- Take 2 minutes
- What words or phrases describe your students (pick two)
  - Undergraduate
  - Graduate
- Share with your table

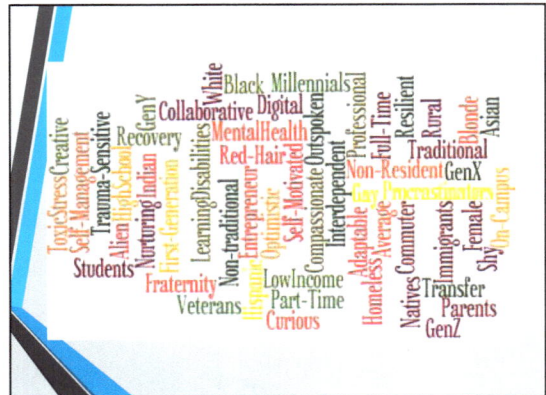

Cell Phone Use (Pew, 2018)

- 95% all American adults use cell phones (77% smart phones)

- Racial/ Ethnic
-

	Any Cell	Smart	Non-Smart
White	94%	77%	17%
Black	98%	75%	23%
Hispanic	97%	77%	20%

Other Devices (Pew, 2018)

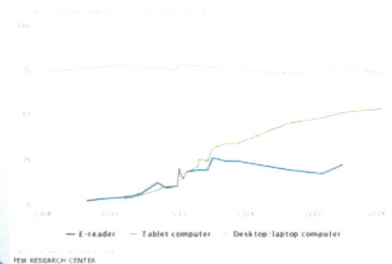

E-reader — Tablet computer — Desktop/laptop computer

PEW RESEARCH CENTER

Smart Phone Use (Pew, 2018)

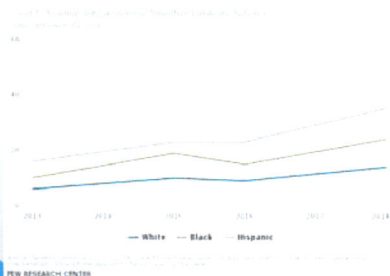

— White — Black — Hispanic

PEW RESEARCH CENTER

5 generations in the workforce (Shah, 2015)

Differences

- *How and where did Kennedy die?*

- Assassination in Dallas, TX – Traditionalist, Baby Boomer
- Plane crash near Martha's Vineyard, MA – Gen X
- Kennedy who? – Gen Y, Gen Z

Generations: 2020 Projection

2020: National Projection

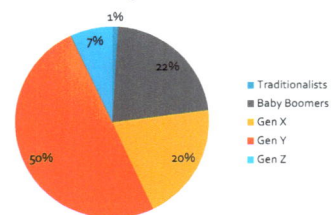

- Traditionalists
- Baby Boomers
- Gen X
- Gen Y
- Gen Z

Source: Future Workplace Survey; Mentoring Millennials (Meister & Willyerd)

Digitally-Connected

Generational Differences x Cultural Diversity

- Increasing scholarship focusing on the interaction between generational differences and cultural differences in the classroom.

- College classroom flip; gender
- World populations (China, India, U.S.)
- Languages, multi-lingual
- Glass, concrete ceilings (80/100)

Technology-free or Technology-friendly?

- How do you use technology (cell phones) in the classroom?

Technology may be used to

- Provide multimodal learning strategies (e.g., aural, visual, kinesthetic)
- Enhance communication style
- Increase technical skills
- Validate students' diversity
- Nurture a sense of community

Examples

- Powtoon – animated presentations
- Progeny – free online pedigree tool
- Qualtrics – online survey tool
- Twitter

Technology

- This employee is a TRUE digital native; technology is easy to use; instinctive use; adaptive to platforms, technology

- Traditionalists (1922-1945)
- Baby Boomers (1946-1964)
- Generation X (1965-1980)
- Generation Y, Millennials (1981-2000)
- Generation Z, Gen 2020, iGen, Post-millennials (2001-present)

"Students do not need a teacher, they have 'Google'"

What do Students need?

- Validate students' diversity
- Encourage students' strengths
- Motivate students' needs
- Nurture a sense of community

Poor Kids (2012)

BANNED
UNK library:
Films on Demand

DIVERSITY ISSUES

White/ European American

Strengths	Challenges
Commitment to Family	Balancing work and family
Enjoyable time together	Maintaining physical and emotional health
Ability to cope with stress and crisis	Creating healthy relationships in a society that glories winning, money, and things
Spiritual well-being	Learning about cultures; being sensitive
Positive communication	Preserving natural environment; consumption
Appreciation and affection	

Poor Kids,
2012 and 2017

Student's traumatic past

African American

Strengths	Challenges
Strong kinship bonds	Judged a financial risk
Strong work orientation	Feeling powerless
Flexibility in family roles	Building self-esteem
Strong motivation to achieve	Facing high male mortality
Strong religious orientation	Overcoming discrimination
Caring parenting	Achieving higher levels of education
Egalitarian marriages	Intraracial violence
	Identifying male role models

American Indian/ Native American

Strengths	Challenges
Extended-family system	Conflicting values of tribe with U.S. society
Traditional beliefs	Maintaining family traditions
High family cohesion	Staying cohesive and connected
Respect for elders	Identifying role models
Bilingual language skills	Achieving higher levels of education
Tribal support system	

Asian American

Strengths	Challenges
Strong family orientation	A need to relax personal expectations
Filial piety (respect for elderly)	Maintaining ties with kin
High value on education	Overcoming emotional vulnerability
Well-discipline children	Overcoming the stigma of seeking help
Extended-family support	Trusting those outside the group
Family loyalty	Relaxing the focus on work

Distance

- Technology for interviews
- - phone
- Facetime, Skype
- Time difference

Hispanic / Latino American

Strengths	Challenges
Familialism	Remaining family centered
High family cohesion	Maintaining traditions
High family flexibility	Gaining financial resources
Supportive kin network	Language barriers
Equalitarian decision making	Overcoming social and economic discrimination
Strong ethnic identity	Relocation issues
	Achieving higher levels of education
	Acculturating across generations

Language

- FAMS 450, Elder Life Research Projects
- Allow of distance
- Allow for language differences ; video in Spanish
- Cultural (religious, family traditions); image "captured"

Identify one person in different generation; different cultural group

Traditionalists Silent Veterans 1922-1945	Baby Boomers 1946-1964	Generation X 1965-1980	Generation Y Millennials 1981-2000	Generation Z Gen 2020 iGen Post-millennials 2001-present

Overview

- Technology use and adaption
- Generational differences
- Cultural considerations
- University e-learning

- *there is little research on FIVE generations of workers and even less on five generations of learners*

THANK YOU!

Research Compliance in the Cloud

Bryan Fitzgerald (UNCA)
Bryan Kinnan (UNCA)

ITAR, CUI, NIS 800-171, and NIST 800-53 are all standards that the University of Nebraska was looking to become compliant within the summer of 2016. After speaking with partners at Microsoft about the University's goals, needs and objectives, it turns out that such a space does exist in Microsoft's governmental o365 space. The goal with this presentation is to quickly highlight the goals, needs, objectives that were defined as part of this project with a look back at the timeline, costs and the pros and cons of going in this direction from the University perspective. The session will conclude with an open discussion with the audience about policies, procedure and next steps, including the roadmap and areas of improvement.

This presentation featured:

- the steps necessary to obtain a GCC High environment license.
- the compliance needs that are met through this cloud option.
- the difference between the educational option and the compliance option.

Research Compliance in the Cloud

2019 Innovation in Pedagogy & Technology Symposium

Nebraska

Agenda

▸ ▸ ▸

- Rules, rules, and more rules…
- On and off campus?
- So what's the catch?
- Okay, tell me more.
- What's next?
- Q&A

Nebraska

Compliance…

▸ ▸ ▸

Nebraska

It is where?

▸ ▸ ▸

	Commercial Cloud	GCC	GCC High
Customer Eligibility	Any customer	Federal, SLG, Native American Tribes, Contractors[1]	Federal, Contractors[1]
Data Residency	US and International	Continental US	Continental US – US NAT support only
Accreditation	FedRAMP Moderate ATO	FedRAMP Moderate ATO, DOD SRG L2	DISA FedRAMP + ATO, DOD SRG L4[3]
Other Relevant Controls	SAS, ISO, HIPAA, and others	CJIS, IRS 1075, NIST 800-53r4	NIST 800-171, NIST 800-53r4[4]
ITAR Support	No	Significant customer requirements[2]	Yes
Network Connectivity	Express Route or Internet	Express Route or Internet	Express Route or Internet
Azure Dependency	Azure (public)	Azure (public)	Azure Government

Nebraska

The requirements are?

▶ ▶ ▶

Microsoft Personnel Screening and Background Checks	Description
U.S. Citizenship	Verification of U.S. citizenship
Employment History Check	Verification of seven (7) year employment history
Education Verification	Verification of highest degree attained
Social Security Number (SSN) Search	Verification that the provided SSN is valid
Criminal History Check	A seven (7) year criminal record check for felony and misdemeanor offences at the state, county, and local level and at the federal level
Office of Foreign Assets Control List (OFAC)	Validation against the Department of Treasury list of groups with whom U.S. persons are not allowed to engage in trade or financial transactions
Bureau of Industry and Security List (BIS)	Validation against the Department of Commerce list of individuals and entities barred from engaging in export activities
Office of Defense Trade Controls Debarred Persons List (DDTC)	Validation against the Department of State list of individuals and entities barred from engaging in export activities related to the defense industry
Fingerprinting Check	Fingerprint background check against FBI databases
Department of Defense IT-2	Staff requesting elevated permissions to customer data or privileged administrative access to Dept of Defense SRG US service capacities must pass Department of Defense IT-2 adjudication based on a successful OPM Tier 3 investigation

[1] Applies only to personnel with temporary or standing access to customer content hosted in Office 365 US GCC-High or DOD clouds

So what do I get?

▶ ▶ ▶

Office 365 Services	Office 365 US Government G1	Office 365 US Government G3	Office 365 US Government G5
Office Online	Yes	Yes	Yes
Office M365 ProPlus	No	Yes	Yes
Exchange Online	Yes	Yes	Yes
Exchange Online Protection	Yes	Yes	Yes
SharePoint Online	Yes	Yes	Yes
OneDrive for Business	Yes	Yes	Yes
Skype for Business (Instant Messaging & Presence)	Yes	Yes	Yes
Voice – Phone System, Audio Conferencing	No	No	Yes
Power & Fax	No	No	No
Project Online	No	No	No
Visio Online	No	No	No
Yammer Enterprise	No	No	No

Is that it?

▶ ▶ ▶

Office 365 Suite Features	Office 365 US Government G1	Office 365 US Government G3	Office 365 US Government G5
Microsoft Bookings	No	No	No
Microsoft Flow	Yes	Yes	Yes
Microsoft Forms	Yes	Yes	Yes
Microsoft Graph API	Yes	Yes	Yes
Microsoft MyAnalytics	No	No	Yes
Microsoft Planner	Yes	Yes	Yes
Microsoft PowerApps	Yes	Yes	Yes
Microsoft StaffHub	No	No	No
Microsoft Stream	Yes	Yes	Yes
Microsoft Sway	No	No	No
Microsoft Teams	Yes	Yes	Yes
Office Delve	Yes	Yes	Yes
Office 365 Groups	Yes	Yes	Yes
Microsoft Stream	Yes	Yes	Yes

In English, please!

▶ ▶ ▶

Office 365 Suite Features	Office 365 US Government G1	Office 365 US Government G3	Office 365 US Government G5
Microsoft Bookings	No	No	No
Microsoft Flow	Yes	Yes	Yes
Microsoft Forms	Yes	Yes	Yes
Microsoft Graph API	Yes	Yes	Yes
Microsoft MyAnalytics	No	No	Yes
Microsoft Planner	Yes	Yes	Yes
Microsoft PowerApps	Yes	Yes	Yes
Microsoft StaffHub	No	No	No
Microsoft Stream	Yes	Yes	Yes
Microsoft Sway	No	No	No
Microsoft Teams	Yes	Yes	Yes
Office Delve	Yes	Yes	Yes
Office 365 Groups	Yes	Yes	Yes
Microsoft Stream	Yes	Yes	Yes

It must be complicated, right?

▶ ▶ ▶

- Demo Time!

What's next?

▶ ▶ ▶

- InTune
- Password Self-Service
- Policies
- Teams
- Encryption

Q&A

▶ ▶ ▶

- Questions and Answers
- Comments?

Our Contacts

Bryan Fitzgerald – bfitzgerald@Nebraska.edu
Bryan Kinnan – bkinnan@Nebraska.edu

Academic Integrity in Higher Education

Tareq Daher, Ph.D. (UNL)
Tawnya Means, Ph.D. (UNL)

Did you know that a 2017 survey from Kessler International has found that 86% of students report that they have cheated in school and 97% of those students report that they have never been caught? Given these shockingly high statistics, what are you doing to deter cheating in your classes? This session will present what you need to know about academic integrity in higher education. We will discuss why and how students cheat in higher educational settings, strategies to manage student expectations and perceptions to encourage them to uphold the highest standards of academic integrity, discuss how to deter cheating and provide communication practices that you can easily implement in your classes.

This presentation featured:

- The reasons of why and how students cheat.
- deterrence methods, tools and tips.
- communication messages to encourage student integrity.

Three types of tendencies

1. Would never knowingly cheat

2. Those who "rationalize"

3. Opportunity to succeed

How do students cheat?
1. Paid homework and papers

How do students cheat?
1. Paid homework and papers

How do students cheat?
2. Sharing and getting materials from others

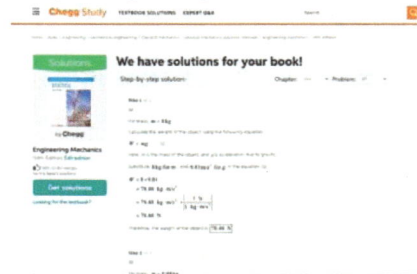

How do students cheat?
3. Copying answers from communities

"Today I graduated and I couldn't have done it without God and Quizlet"

"Shout out to Quizlet for making this possible"

https://quizlet.com/

How do we prevent academic dishonesty?

Manage expectations and perceptions

1. Protect the value of **your** degree (students)
2. Let students know it is important to you
3. Show real world application of course content
4. Care about the individual student's success

How do we prevent academic dishonesty?

Deterrence

1. Plagiarism detection
2. Assessment development
3. Proctoring
4. Consequences
5. Communicate

Deterrence

1. Plagiarism detection - Turnitin

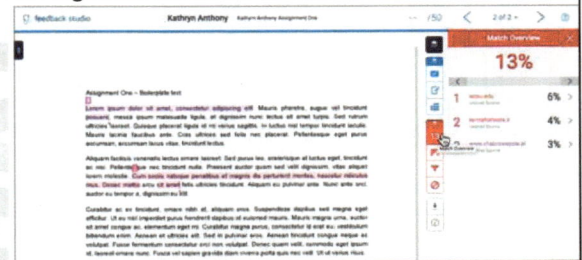

Deterrence

2. Assessment Development

A. Make the assessment mirror reality
B. Cycle through doing and reflection
C. Require transformation
D. Personalize assessments

Deterrence

D. Personalize Assessments

a) Formula-based, open response questions
b) Use real or simulated data
c) Increase frequency
d) Lower stakes
e) Build on previous assessments (projects/tasks)
f) Prepare students (practice, repeated recall)

Deterrence

3. Proctoring

Digital Learning Center – Exam Commons

Deterrence

4. Consequences

In this class, if you are found responsible for academic dishonesty, any of the following action(s) can be imposed:
• give out a warning
• reduce the grade on the assignment, paper, quiz, or exam
• reduce the student's final grade in the course
• assign a zero on the assignment, paper, quiz, or exam
• Request for an "F" to be on the transcript for the course
• Student can be required to pay for and complete an online Academic Integrity Course
• In severe circumstances: Student can be temporarily separated from the University (suspension) or permanently separated from the University (expulsion). Suspension or expulsion can result in changes to the student's visa status.

Deterrence

4. Consequences

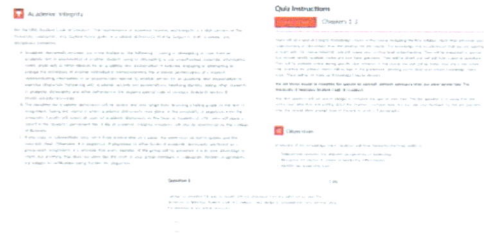

Communicate

Help students see the importance of protecting academic integrity. Some of the messages to share with them include:

When you cheat, you are missing the opportunity to develop a deep understanding of the content. What you didn't learn in one course can be required in the next course or in your future workplace.

Communicate

If grades are not evidence of learning, jobs are filled by incompetent people (cheating the economy)

Communicate

Help students see the importance of protecting academic integrity. Some of the messages to share with them include:
• *Your behavior affects others – Cheating is not just a violation of academic conduct, it is a violation toward a community.*

Communicate

If students are taking government loans and grants, taxpayers dollars are cheated

Communicate

The research is clear. Students who cheat in college will cheat when faced with temptation and opportunity. Cheating will set you up for cheating in life.

Communicate

Cheating puts your personal integrity at risk. What kind of human being do you wish to make of yourself?

Raise Awareness

- https://justdomyhomework.com/
- https://takemyonlineclass.com/
- https://www.studymode.com/
- https://www.coursehero.com
- https://quizlet.com/
- https://edusson.com/
- https://www.noneedtostudy.com/

Raise Awareness

- Participate in the Academic Integrity process at UNL
- Engage in the Ethics and Integrity community of practice – a discussion group of faculty

- Sign up for more information: http://bit.ly/2VTNELU

Featured Extended Presentation

Emerging Technology Trends:
Virtual Reality & Artificial Intelligence

Bryan Alexander, Ph.D.

This session will focus on the two major technological forces likely to impact online education in this extended session. We will explore the possibilities of virtual reality (VR), examining its current pedagogical aspects and institutional support structures. We will also examine upcoming virtual reality uses, including social VR as well as VR as an alternative to videoconferencing. Secondly, we will address artificial intelligence (AI) by reflecting on its current and emerging uses in higher education, from chatbots to AI-powered teaching software and student data analytics. We will continue on a prospective note, looking at trendlines and the history of technology to anticipate medium-term uses of AI in academia. Our conclusion steps back to consider how online education could change as AI starts transforming the labor market and society as a whole.

This presentation featured:

- the impact of VR and AI on higher education.
- the pedagogical aspects of VR.
- how AI-powered technology can impact the student experience.

Plugging into Student Support Services for Student Success

Victoria Brown, Ph.D. (Florida Atlantic)

More students are pursuing their academic career 100% online. As institutions continue to develop strategies in how to create a successful online experience, students still report feelings of isolation. These feelings persist even as faculty and instructional designers include community-building activities and use instructional strategies to engage students with the instructor and each other. Could the missing component be the services beyond the classroom? Attend this session to gather ideas on how to promote engagement through support services in your department or unit.

This presentation featured:
- how to promote engagement through student services in your department or unit.
- the OLC Quality Scorecard for Online Student Support

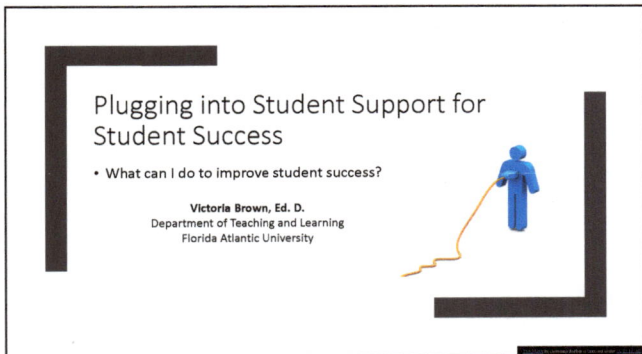

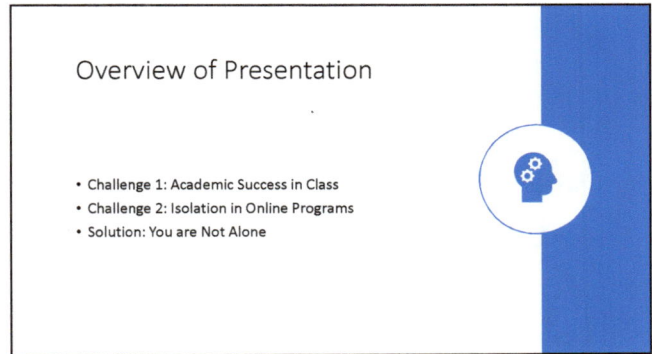

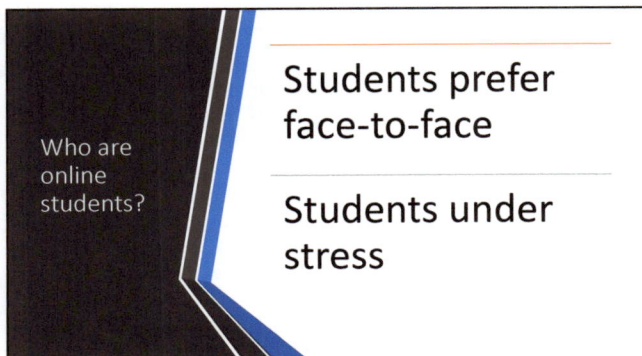

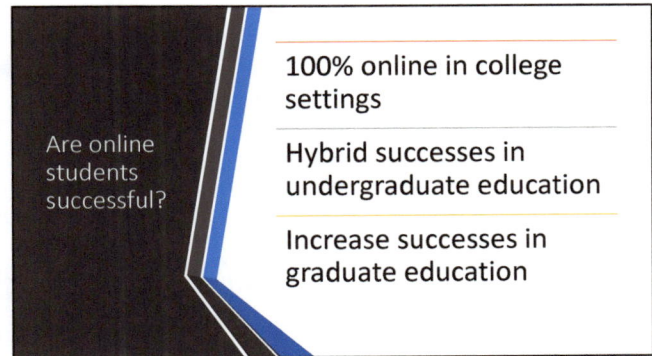

Faculty first success advocates

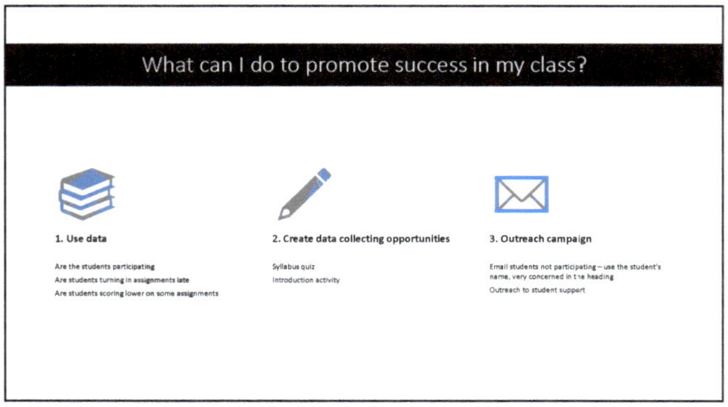

What can I do to promote success in my class?

1. Use data

Are the students participating
Are students turning in assignments late
Are students scoring lower on some assignments

2. Create data collecting opportunities

Syllabus quiz
Introduction activity

3. Outreach campaign

Email students not participating – use the student's name, very concerned in the heading
Outreach to student support

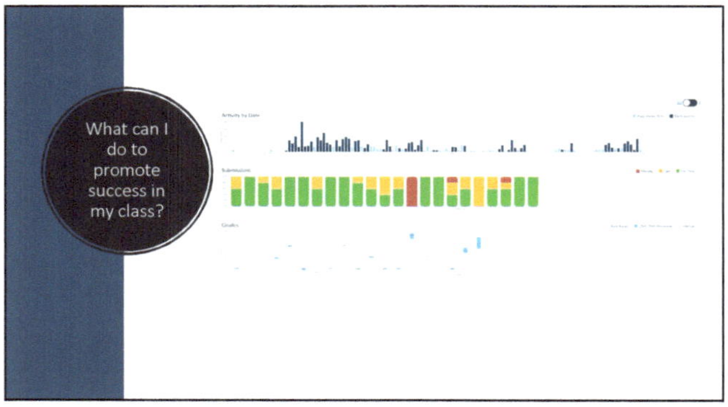

What can I do to promote success in my class?

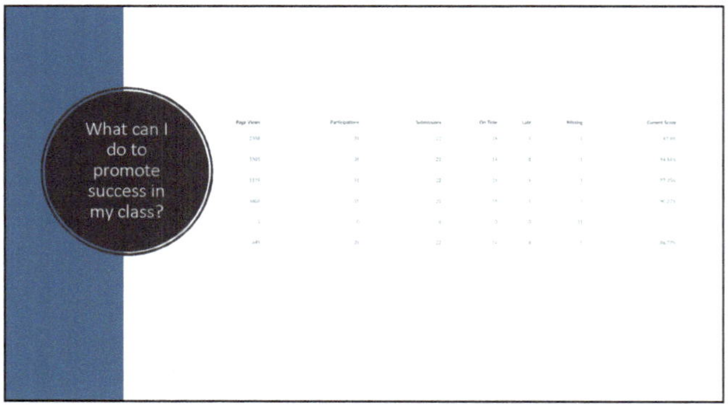

What can I do to promote success in my class?

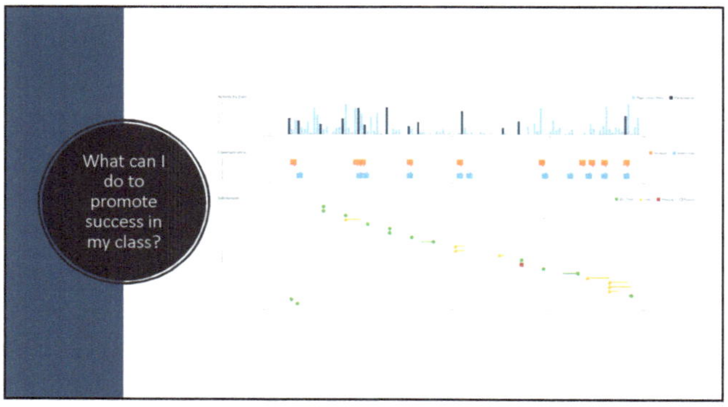

What can I do to promote success in my class?

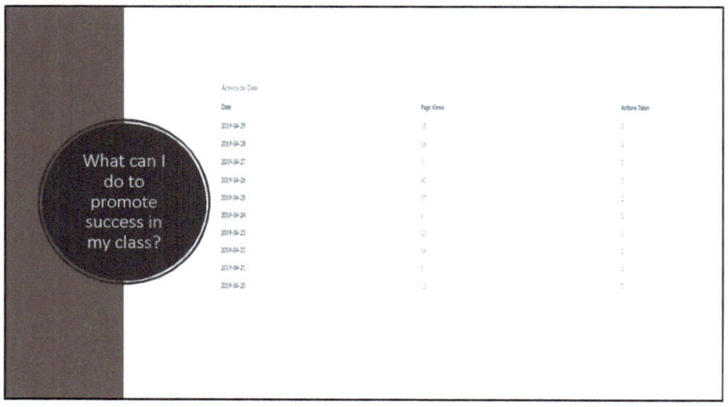

What can I do to promote success in my class?

What can I do if a student is struggling?

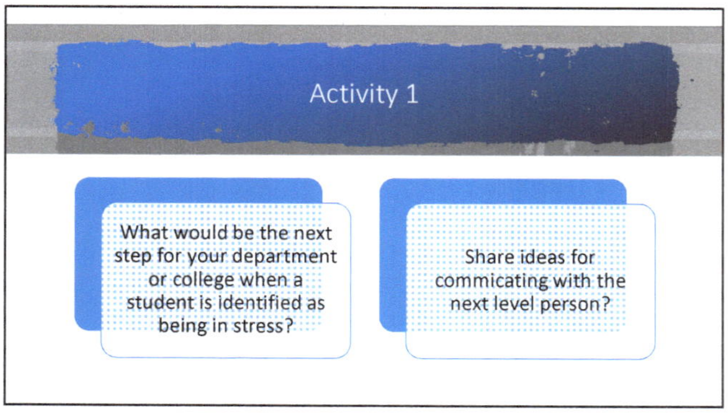

Activity 1

What would be the next step for your department or college when a student is identified as being in stress?

Share ideas for commicating with the next level person?

Satisfaction with online learning

Students rate engaging classes higher

Students report isolation

Education is a social experience.

Training is teaching specific skills.

Create an online educational experience. Plug-in to student services and create social interaction beyond the class.

Create an Online Community

This Photo by Unknown Author is licensed under CC BY-ND

Student Life Cycle

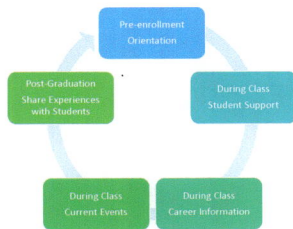

Pre-enrollment Orientation

Post-Graduation Share Experiences with Students

During Class Student Support

During Class Current Events

During Class Career Information

Activity 2

Search website for potential services

Search website for how to recommend students for services

References

1. Education Advisory Board. (2013). The path to persistence: Strategic interventions for adult and online learners.
2. Florida State University Steering and Implementation Committee for 2025 Online Education. (2016). Cost of online education. Retrieved from https://www.flbog.edu/board/advisorygroups/_doc/online/ADDENDUM_02a1_Affordability.pdf
3. Jones, P. R. (2016). The structure and pedagogical style of the virtual developmental education classroom: Benefit or barrier to the developmental learning process? *International Journal of Language and Literature, 4*(43-48.
4. Ke, F. & Kwak, D. (2013). Online learning across ethnicity and age: A study on learning participation, perception, and learning satisfaction. *Computers & Education, 61*, 43-51.
5. Protopsaltis, S. & Baum, S. (2019). Does online education live up to its promise? A look at the evidence and implications for federal policy. Retrieved from https://mason.gmu.edu/~sprotops/OnlineEd.pdf

Adapting to the Changing Needs of Students: A Collaborative Approach to Programmatic Change

Amber Alexander (UNK)

Doug Biggs, Ph.D. (UNK)

Steve McGahan (UNK)

Establishing a quality online program takes years, but the changing face of student populations requires that these programs adapt the structure and curriculum to reflect change over time. As student population and discipline requirements change, programs must change as well to ensure that graduates leave the program with the discipline, knowledge and secondary skills that will facilitate success in their careers. This presentation will look at the challenges and strategies that have been used by the UNK M.A. in History program to continue to serve the student population. Facets of program pedagogy, faculty buy-in, coordination of student progress, an introduction of new faculty and instructional design will be addressed as presenters share their insights into the process of adapting an already established program to the changing needs of the student population.

This presentation featured:

- how to develop a collaborative team to address changing programmatic needs.
- the current and future needs of an established program in a changing field.
- examples of adapting student learning and skills while maintaining course and program integrity.

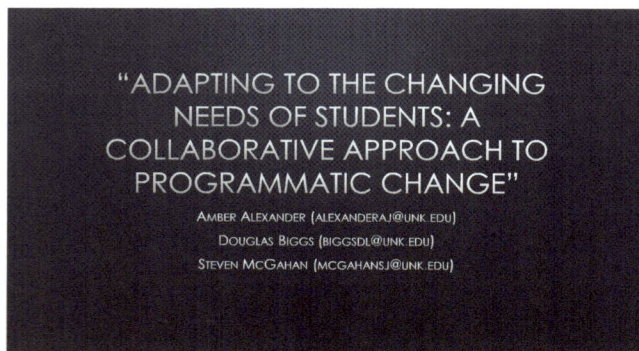

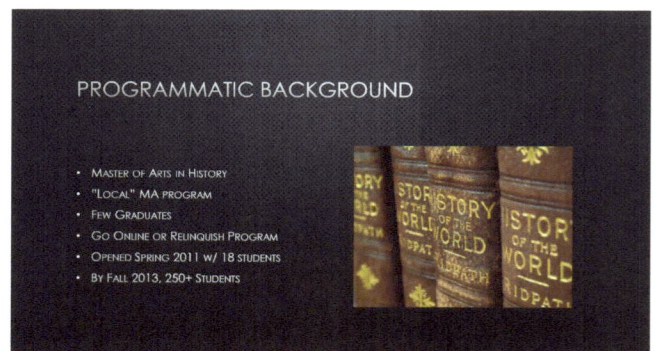

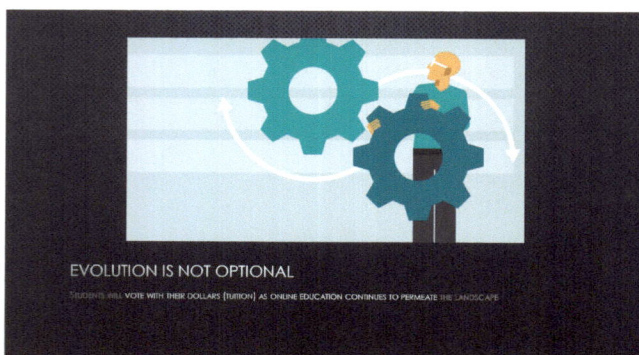

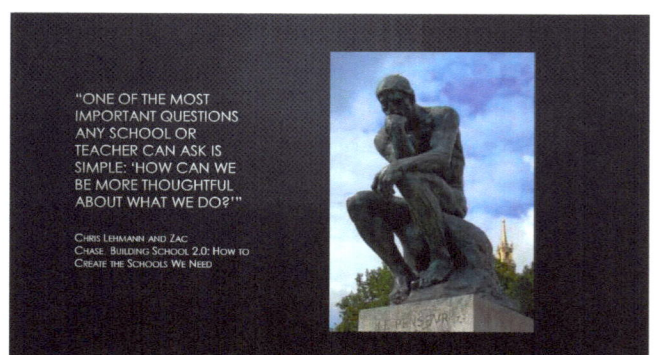

COOPERATIVE TEAM

FACULTY
COORDINATOR
INSTRUCTIONAL
DESIGNER

Empathize Define Ideate Prototype Test

EMPATHIZE
FIND THE PAIN POINTS THAT ARE HINDERING STUDENT SUCCESS

THE CHANGING FACE OF GRADUATE STUDENTS

- FEWER TRADITIONAL STUDENTS
- LONGER MATRICULATION TIMES
- MORE WORKING PROFESSIONALS

OUR STUDENTS ARE LISTENING

- ONLINE PROGRAMS CREATE COMPETITION NATIONALLY INSTEAD OF REGIONALLY
- PROGRAMS MUST EVOLVE TO MAINTAIN VIABILITY WITH STUDENTS WHO ARE "SHOPPING" FOR THEIR DEGREES

THE ONLINE PARADIGM

STUDENTS FUNCTION IN A WORLD WHERE FACTS ARE AT THEIR FINGERTIPS AND EMPLOYERS CAN KNOW THEM BEFORE THE INTERVIEW

searching...

DEFINE
A STUDY OF THE ISSUES THAT SURROUND THE INITIAL IDENTIFIED PROBLEMS

THE CURRENT PROGRAM

- 36 CREDIT HOUR PROGRAM
- THESIS TRACK (C. 55%)
- NON-THESIS TRACK (C. 45%)
- "NEED" FOR A "SUMMATIVE ASSESSMENT" IN BOTH TRACKS
- COMPREHENSIVE EXAMS

THE PROBLEM: COMPREHENSIVE EXAMS

- UNSUCCESSFUL MEASURE
- ASSESSMENT RUBRIC ISSUES
- STUDENTS & FACULTY HAD ISSUES
  - PERFORMANCE ISSUES
  - NO DELIVERABLE PRODUCT, UNLIKE THESIS STUDENTS

OTHER IDENTIFIED ISSUES

- HISTORY FACULTY BUY-IN
- TECHNOLOGY ISSUES
- DEVELOPMENT TIME

IDEATE

BRAINSTORM IDEAS TO SOLVE THE IDENTIFIED ISSUE(S)

IDENTIFYING OPTIONS

- THE "TEAM" BEGAN DISCUSSIONS ABOUT WHAT COULD REPLACE THE COMPREHENSIVE EXAM
- CRITERIA WERE ESTABLISHED
- DIFFERENT IDEAS WERE INVESTIGATED

THE IDENTIFIED SOLUTION: DIGITAL PORTFOLIO

- ONLINE PORTFOLIO OF WORK BASED ON CURRICULUM AND CONTENT EMPHASIS
- COMPONENTS COLLECTED THROUGHOUT THE DEGREE
- CREATE A ROADMAP OF LEARNING AND SKILLS EVOLUTION
- STATEMENT OF PURPOSE

PROTOTYPE

SELECT AND DEVELOP THE MOST PROMISING IDEAS INTO A FUNCTIONAL SOLUTION

FACULTY BUY-IN

- CERTAIN FACULTY WERE RESISTANT TO THE CHANGE
- THE VETTED METHODOLOGY WAS VIEWED AS THE CORRECT PATH (STATUS QUO)
- FACULTY APPROVAL
- GRADUATE COUNCIL

BENEFITS OF ONLINE PORTFOLIO VS. EXAM

- AN ARTIFACT STUDENTS COULD TAKE WITH THEM AND BUILD ON IN THE FUTURE
- MORE IN LINE WITH CURRICULUM
- ELIMINATE PROCTORING ISSUES
- BETTER ASSESSMENT?

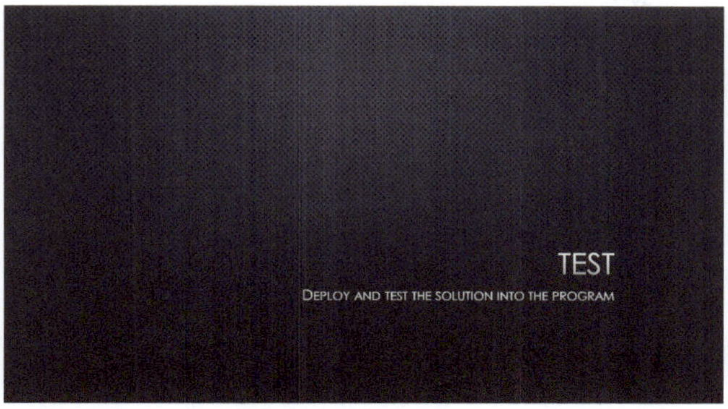

TEST
Deploy and test the solution into the program

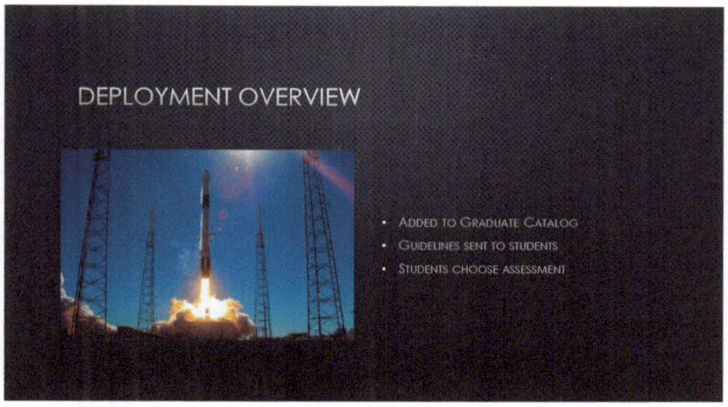

DEPLOYMENT OVERVIEW
• Added to Graduate Catalog
• Guidelines sent to students
• Students choose assessment

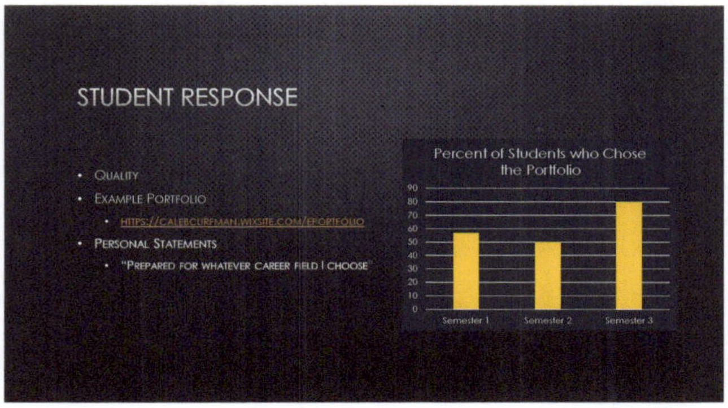

STUDENT RESPONSE
• Quality
• Example Portfolio
• https://calebcurfman.wixsite.com/eportfolio
• Personal Statements
• "Prepared for whatever career field I choose"
Percent of Students who Chose the Portfolio
90
80
70
60
50
40
30
20
10
0
Semester 1
Semester 2
Semester 3

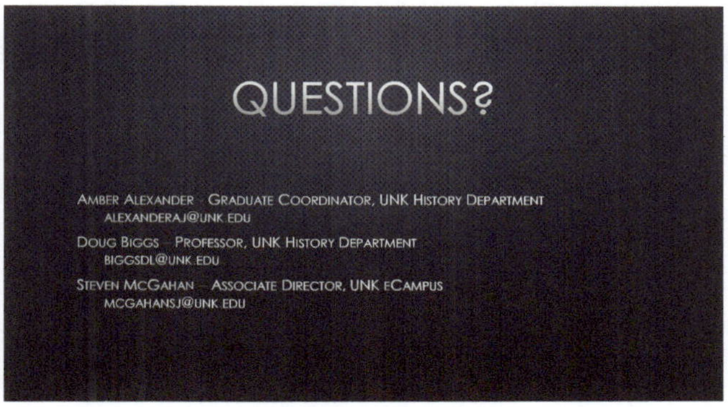

QUESTIONS?
Amber Alexander – Graduate Coordinator, UNK History Department
alexanderaj@unk.edu
Doug Biggs – Professor, UNK History Department
biggsdl@unk.edu
Steven McGahan – Associate Director, UNK eCampus
mcgahansj@unk.edu

Cybersecurity Escape Room Challenge - Version 2

Cheryl O'Dell (UNCA)

Do you have what it takes to solve the puzzles and find all the information to "escape" the presentation in 45 minutes? Don't worry - you will have help, and we will be asked to leave the room for the next presentation anyway! Come participate in this unique experience and learn some security awareness at the same time. This wildly successful effort is a presentation to build teamwork skills, utilize everyone's unique talents and have some fun in the process of learning about cybersecurity. In the fall of 2018, we thwarted Dale Isa Spy's attempt of stealing university data. In the spring of 2019, a new scenario and new puzzles await those who dare to try the Cybersecurity Escape Room Challenge.

This presentation featured:

- security surrounding two-factor authentication, WiFi and using only secure websites.

Slide 1

Another goal – learn about web site security.

Browsing Smart = Browsing Safe
Don't rely on your browser to protect you

https://www

Your browser can't stop you from visiting a dangerous site or downloading malicious software.

Surf smart to stay safe!

Remember these best practices:
- Do not click links or downloads in pop up windows
- Avoid the lure of free content — there's almost always a catch
- Be cautious of shortened URLs
- Turn off "Auto Complete" and "Remember Me" features

Nebraska

Slide 2

Rules

- You have 45 minutes
- Everything you need to solve the problems are in the kit
- The process is not linear – meaning:
  - You don't have to solve puzzles in order. The current puzzle(s) may need help from future discoveries.
  - If you are stuck, maybe you have not found all the clues.
  - Multiple puzzles can be worked on simultaneously.
  - Subgroups may work on different puzzles.
- We went over two handouts – but they still are clues to solving puzzles!

Nebraska

Slide 3

Ready, Set, Escape

Any questions before we start?

https://youtu.be/FCO95esUzBw

Nebraska

Online Course Design 101

Jena Asgarpoor, Ph.D. (UNL)

Anchored by 24 years of online course development and delivery, the author draws upon that experience as well as Backward Design and design strategies promoted by Quality Matters Organization to highlight critical issues in development of an online class. Audience in this session will learn about design considerations that make an online course a positive, engaging and successful learning experience for the learners. Choosing and developing the appropriate resources, assignments, discussion activity and projects is covered, and the importance of promoting connectedness and sense of community through three types of interactions (learner-to-learner, learner-to-content and learner-to-faculty) is emphasized. Using Backward Design, the point will be made that choosing the right resources and activities is critical in the development stage within any delivery format and online is no exception. Examples will be provided in the context of a fully online, 8-week Canvas course in Engineering Leadership, which is an elective in the Master of Engineering Management program in the College of Engineering.

This presentation featured:

- strategies to improve your syllabus, course navigation, content/activities and interactions/engagements.
- Backward Design and how it can be utilized in a fully online course.
- the correct resources and activities in order to better the delivery format of your course.

Design for Teaching Effectiveness

Course Portfolio and Peer Review of Teaching (PRT) Project
Year-long project
Three Benchmark Memos and Analysis

Describe course and its goals → Describe course activities → Document and analyze learning

Bernstein, Burnett, Goodburn, and Savory (2006). *Making Teaching & Learning Visible (Jossey-Bass)*

Effective Online Course Design

Quality Matters (QM) ®

- Best practices: online/blended Design

- Rubric to evaluate courses

- Certification to use rubric & to review

- Backward Design Philosophy

Assessments

Instructional Materials | Learning Activities | Course Tools & Technologies

Learning Objectives

Source: Quality Matters

Quality Matters General Standards

1. Design and navigation

2. Learning Objectives
3. Assessment
4. Instructional Materials
5. Learning activities and interactions
6. Course technology

7. Learner support
8. Accessibility

Quality Matters Standards

Standard 1: Overall course design is made clear to the learner at the beginning of the course.

1.1: Instructions make clear **how to get started** and where to find various course components.

1.2: Learners are introduced to the **purpose and structure** of the course.

1.3: Communication **expectations** for online **discussions, email**, and other forms of **interaction** are clearly stated.

1.4: Course and **institutional policies** with which the learner is expected to comply are clearly stated within the course, or a **link to current policies** is provided.

1.5: **Minimum technology** requirements for the course are clearly stated, and information on how to obtain the technologies is provided.

1.6: **Computer skills** and digital information literacy skills expected of the learner are clearly stated.

1.7: Expectations for **prerequisite knowledge** in the discipline and/or any required competencies are clearly stated.

1.8: The **self-introduction by the instructor** is professional and is available online.

1.9: **Learners** are asked to introduce themselves to the class.

Ease of Navigation and Locating Information

- Online education can be stressful to a new student

- Even seasoned OL students may have stress since design and navigation may differ for each course

- At the program level, a course template would be helpful for consistency

---

Orientation Module

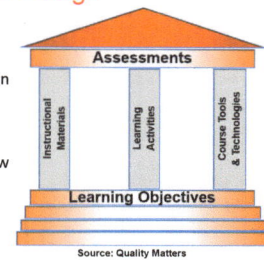

Orientation - Tour de Class Video

- *Tour de Class* video (10 – 15 minute) in each course
  - Save students exploration time
  - Guide them in navigating the online course
  - Give overview of syllabus & course requirements
  - Explain what to expect each week & where to find it

- Covered in Tour de Class video:
  - Syllabus content
  - Activity to acknowledge syllabus & course policies
  - Set up notifications to receive emails
  - *Introductions* Discussion area

Orientation – Syllabus & Acknowledgement

- Central location for policies, expectations, outline

- Summary table for each module and assignments

- Syllabus area in Canvas and Due Dates

- Syllabus *acknowledgement* activity
  - No credit assignment
  - Single-question quiz to submit
  - Probably more appropriate for UG students; but I use in all classes
  - Due date in the first few days of class to gain access to homework

---

Introductions: Build Community https://use.vg/45YzGJ

- Professor's welcome note (video message)
  - VidGrid: A "must" for online teaching

- Make it personal (experience, family, hobbies)
  - Share a fun fact or two about yourself

- Give assurance that you are accessible
  - Email
  - Phone (I share my personal phone number)
  - Zoom (one-on-one)
  - Zoom (group session, recorded)
  - Online Office Discussion Area – Use it instead of announcements to promote DB use

- Ask students to also share a similar video
  - Tell them to share as much as their comfort level allows

Nebraska
Lincoln

Course Navigation: Student view

- Minimum course areas
- Integrate links into modules for one-stop interaction with content
  Hide:
  - Files
  - Pages
  - Quizzes

- Module content:
  - Description of content
  - Learning goals
  - Content & learning materials
  - Discussion board activity & assignments

https://canvas.unl.edu/courses/58950

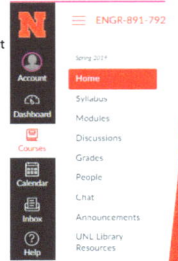

Nebraska
Lincoln

Demo:
- Screen capture video with VidGrid
- Quiz and Comment functions
- https://use.vg/5TKpSZ

Nebraska
Lincoln

Design: Build a cohesive community of learners
Improve retention, persistence, and success

- Clayton Christensen's quote about disruptive innovation (author of 2011 book, *The Innovative University*)
  - Anything beyond the 10th row in a large lecture hall is distance education, anyway!
- Build community & engage students: Have a strong *Online Presence*
- Communicate frequently:
  - Set expectations: Tell them what happens if they miss their DB deadline
  - All-class messages (inbox and DB): Reminders, Encouraging notes, Hints on assignments, etc.
  - Targeted messages using Canvas grade center – for both types of performers
  - Move recent Modules & Discussions to be on top
  - Give timely response/feedback (stick to your syllabus promise; Mine is 48 hrs & weekly, respectively)
- Avoid instruction by email to build community: **Design for it**
  - Online Office DB (Administrative issues)
  - Weekly ungraded discussion area
  - Graded discussion (Thursdays/Sundays with 2 replies)
  - Introductions area: respond to each person with video
  - Encourage updating profile with a personal picture (give instructions)
- Interaction: Schedule Zoom Group Sessions (record for asynchronous viewing)
- Give useful feedback on assignments to each student (in addition to rubric)
- Incorporate audio/video in your feedback
- Encourage posting their weekly discussions and replies as a video file
- Give a mid-term survey

Quality Matters: Standards 2 - 6

Standard 2 – Learning objectives describe what learners will be able to do upon completion of the course. They establish a foundation upon which the rest of the course is based.

Standard 3 – Assessments are integral to the learning process; designed to evaluate the learner's progress in achieving the learning objectives.

Standard 4 – Instructional materials enable learners to achieve stated learning objectives.

Standard 5 – Learning Activities and Learner Interaction: Learning activities facilitate and support learner interaction and engagement.

Standard 6 – Course Technology: Course technologies support learners' achievement of course objectives or competencies.

Nebraska
Lincoln

This is me in mid-August 2017, when I joined the MEM!

I had 6 weeks to develop my first course

The following is something that I would have talked about in my presentation. But, I have created it in text form here since I will be emailing my slides to you.
1. Temptation was high to just grab a book and build a syllabus around it (:-o) Quite likely, students would not really know the difference if I were to do that versus go through the backward design and build the course based on sound teaching and learning practices. As the pictures and the animation on the next slide try to convey, I had a choice to make. I decided to do the right thing! The Angel defeated the Devil ☺
2. The next few slides mention an "interactive" e-book. Towards the end of this video, there is a segment where I quickly demo the book: https://use.vg/37XzGZ
3. In slide 21, where you see L-C, L-F, L-L, the letters stand for: Learner, Faculty, and Content

---

Do the right thing by the students!

Professional graduate program. Students have:
- Engineering UG degree
- ≥ 2 years of engineering work experience

Design Issues
- Rigor/expectations must be at the graduate level
- But, these students have no prior experience or background in the topic
- Must be application oriented – integrate theory to practice
- Provide opportunity for introspection (What does it mean to "ME"?)

Objectives — Describe what learners will be able to do and are relevant to course level
- Fundamentals, general knowledge
- Evaluate and apply
- Introspection and self-discovery

1. Articulate the components of several leadership theories including Transactional and Transformational leadership, Charismatic leadership, models of power and influence, Leader-Member Exchange theory, and Motivation theory.

2. Describe the fundamental differences between management and leadership, and how each concept affects organizational performance.

3. Explain how personality traits, motives, and personal characteristics influence leader behavior.

4. Demonstrate ability to apply leadership theory and research in personal and organizational contexts.

5. Evaluate and integrate personal motives, traits, and characteristics through a reflection summary of one's own personal leadership style.

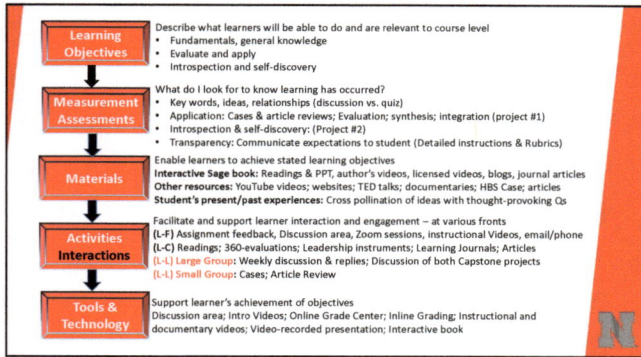

Slide 1

Learning Objectives	Describe what learners will be able to do and are relevant to course level • Fundamentals, general knowledge • Evaluate and apply • Introspection and self-discovery
Measurement Assessments	What do I look for to know learning has occurred? • Key words, ideas, relationships (discussion vs. quiz) • Application: Cases & article reviews; Evaluation; synthesis; integration (project #1) • Introspection & self-discovery: (Project #2) • Transparency: Communicate expectations to student (Detailed instructions & Rubrics)
Materials	Enable learners to achieve stated learning objectives **Interactive Sage book:** Readings & PPT, author's videos, licensed videos, blogs, journal articles **Other resources:** YouTube videos; websites; TED talks; documentaries; HBS Case; articles **Student's present/past experiences:** Cross pollination of ideas with thought-provoking Qs
Activities Interactions	Facilitate and support learner interaction and engagement – at various fronts (L-F) Assignment feedback, Discussion area, Zoom sessions, instructional Videos, email/phone (L-C) Readings; 360-evaluations; Leadership instruments; Learning Journals; Articles (L-L) Large Group: Weekly discussion & replies; Discussion of both Capstone projects (L-L) Small Group: Cases; Article Review
Tools & Technology	Support learner's achievement of objectives Discussion area; Intro Videos; Online Grade Center; Inline Grading; Instructional and documentary videos; Video-recorded presentation; Interactive book

Slide 2

Design: Learning Activities and Materials

Hook activity

Angelo, T and Cross, P (1993). *Classroom Assessment Techniques: A Handbook for College Teachers* (Jossey-Bass)

➢ ***New York Underground*** (Documentary)
  Building NYC's underground rail system

  https://www.youtube.com/watch?v=N0Xm9WssR4w

Slide 3

Design: Learning Activities and Materials

Engage the student with the course & material

Weekly Discussion & Replies
Scenarios - hypotheticals
• You are the leader of a unit that is being merged
• You are leading a team with a qualified member who is not a team-player

Questions - provocative
• Can leaders be more or less authentic? Or, is authenticity like pregnancy: either you are or aren't authentic?

• Defend or refute: Opportunistic leadership can be justified in certain situations.

• What would follower competence and commitment look like in your work unit or organization and how are they conceptualized?

Slide 4

Design: Learning Activities and Materials

Relevant to engineering; reflection & introspection

Article Reviews
➢ Space Shuttle ***Challenger*** Disaster (Documentary & Article)
  Evaluation from perspective of leadership failure
  • Maier, Mark (2002, September). Ten Years After A Major Malfunction.... Reflections on "The *Challenger* Syndrome". *Journal of Management Inquiry*, 11(3), 282-292
  • https://www.youtube.com/watch?v=-O_DMyHdg_M
  • https://www.youtube.com/watch?v=P9LSerNokJk

➢ Leadership in **undergraduate engineering** programs (Article)
  • Rottman, C, Sacks, R, and Reeve, D. (2015). Engineering Leadership: Grounding leadership theory in engineers' professional identities. *Leadership*, 11(3), 351 – 373.

Slide 5

Design: Learning Activities and Materials

Interesting, Motivating, Graduate Level

Textbook choice
➢ Peter Northouse, Leadership: Theory and Practice (Sage)
➢ Based on **theory**; emphasizes & summarizes application & practice
➢ **Interactive:** multi-media Integrated throughout each chapter
  ▪ Author's Videos
  ▪ Licensed Videos
  ▪ Podcasts
  ▪ Sage Journal articles
  ▪ Leadership Instruments

Optional Books
  ▪ Kouzes, J.M. and Posner, B.Z. (2017). The Leadership Challenge: How to Make Extraordinary Things Happen in Organizations (6th ed.), Wiley.

  ▪ Strengths-Based Leadership by Tom Rath (Gallup)

Slide 6

Design: Learning Activities and Materials

Introspection and self-awareness (biases, beliefs, values)

360-evaluations (https://studentleadershipcompetencies.com/evaluations/360-evaluation/)
Self-evaluation AND several completed by coworkers, superiors, etc.
➢ Learning and reasoning
➢ Self-awareness and development
➢ Group dynamics
➢ Interpersonal interactions
➢ Civic responsibility
➢ Communication
➢ Strategic planning
➢ Personal behavior

Slide 7

Design: Learning Activities and Materials

Introspection and self-awareness (biases, beliefs, values)

Leadership Instruments
➢ Fifteen (listed on the next slide)
➢ Abridged version of validated instruments
➢ Scored, interpreted, suggest a roadmap (how to improve)
➢ Some are 360-degree multi-rater
➢ A few are scenario based

Slide 8

Design: Learning Activities and Materials

Trait approach	Leadership Trait Questionnaire
Skills approach	A Skills Inventory
Behavioral Appr.	Leadership Behavior Questionnaire
Situational	Situational Leadership (4)
Path-Goal	Path-Goal Leadership Questionnaire
Leader-Member	LMX 7 Questionnaire
Transformational	MLQ form 5X-short
Authentic	Authentic Leadership self-assessment
Servant	Servant Leadership Questionnaire (SLQ)
Adaptive	360, multi-rater feedback (5)
Psychodynamic	Leadership Archetype Questionnaire
Ethical	Perceived Leader Integrity Scale
Team	Team Excellence & Collaborative Leader
Gender	Gender-Leader Implicit Association test
Culture	Dimensions of Culture Questionnaire

Design: Learning Activities and Materials

Introspection and self-awareness (biases, beliefs, values)

Weekly Learning Journals
Reflection on what you learned from resources
Reflection on what you learned from the Leadership Instruments
How do you plan to use it at work and in your life?

- Childhood experiences
- Work experiences (as a leader and as a follower)
- Current life dynamics
- Etc.

Design: Learning Activities and Materials

Relevant to engineering; apply concepts

Cases
- Trait Approach - Food manufacturing
- Transformational – 50 years old Manufacturing
- Ethical – Hiring decision at a University Research Institute

---

Design: Learning Activities and Materials

- Scaffolding of knowledge, activities, and assignments
- Prepare for course capstone project(s)

Capstone 1: *Endurance* documentary & HBS Case

Ernest Shackleton's 1914 expedition to Antarctica
Resources:
- Kohen, Nancy. (2010). *Leadership in Crisis: Ernest Shackleton and the Epic Voyage of the Endurance,* HBS No. 9-803-127. Boston, HBS Publishing.
  https://www.youtube.com/watch?v=oyQ8HHHXHtc

Required students to:
1. Review course content
2. Apply knowledge to evaluate a leader

Capstone 2: Leadership Self Profile

A 10 – 15 minute video presentation:
1. Report self-discovery (beliefs, biases, values)
2. Strengths & weaknesses
3. How does it shape me as a leader now
4. Who I want to be as a leader
5. How will I get there? Road-map

Required students to:
1. Review of course content
2. Evaluate oneself
3. Integrate new knowledge & self Information
4. Apply it to one's life and work
5. Synthesize into a plan of action

Rubrics

Rubrics

Learning Journal Rubric
You've already rated students with this rubric. Any major changes could affect their assessment results.

Criteria	Ratings				Pts
Description of criterion	10.0 pts Demonstrates in-depth analysis and introspection of one's characteristics and style of leadership as they relate to relevant theories, concepts, and personal experiences	7.5 pts Demonstrates adequate insights into one's characteristics and style of leadership as they relate to relevant theories, concepts, and personal experiences	5.0 pts Demonstrates minimal insights into one's characteristics and style of leadership as they relate to relevant theories, concepts, and personal experiences	2.5 pts Demonstrates no insights into one's characteristics and style of leadership as they relate to relevant theories, concepts, and personal experiences	10.0 pts

Total Points: 10.0

Rubrics

Results
(13 of 13)

34. What are 1 or 2 specific things that helped you learn in this class?

- Discussion boards for assignments.
- Excellent textbook.
- The combination of readings, evaluations, videos, and reference articles. I like approaching course topics from all angles with a multitude of resources to go to.

https://crseval.unl.edu/course_report/index/3/166559?survey_trigger_id=212   4/6

10/28/2019   Individual Course Report - ENGR891 Sec. 701

- Overall I felt the course was really good, the professor worked hard to make the online community feel genuine and seemed like she actually cared about the students. It was a pleasant overall experience and I'd recommend the course to others. I appreciated the timely feedback on my work and the quality of the feedback was great. I really felt like the professor cared that I learned the material.
- I did feel as though the instructor was very prompt in her responses and grading. She was also quite constructive in her feedback and I appreciated the level of detail and insight she gave when grading assignments and participating in discussions. I also appreciated that she was pretty consistent in grading based on the rubric and provided specific feedback on what she needed to see on future assignments to receive full credit. This transparency in the grading was refreshing and appreciated from a student perspective.
- Excellent use of various supplementary articles and videos.
- Quick Feedback and the open forum style helped, especially with this being an online class.
- the text book was great and the videos very powerful
- My group helped me a lot and the videos we watched were interesting and helped me stay focused.
- Instructor feedback was very meaningful and helpful. Great assignments to drive points home.

Results
(13 of 13)

35. What are 1 or 2 specific things that caused a problem with your learning in this class?

- I understand that online courses require us to get a little creative with how we communicate, but I just feel that the forced online discussion is a little silly. Much of the conversations are forced and aren't conducted out of authentic interest in the subject at hand. A possible solution might be online video chats back and forth. For whatever reason conversation seems to be more authentic when it is spoken as opposed to typed.
- My group was problematic. They all participated but one of them was not familiar with English and so every time we had a group project which required writing (all of them) she would do her part and then it would have to be completely re-written. Multiple times this occurred and caused the group to take excessive effort to correct and address her entire section as well as their own.
- Nothing, really.
- I was unprepared for the pace and struggled a few weeks but managed to finish strong with a decent grade (I think).
- nothing
- I am not a "theory" thinker so it was a little harder for me to stay focused on the readings
- N/a

Jena Shafai Asgarpoor
jshafai@unl.edu

Learning Objectives

QM Standard 2 – Learning objectives describe what learners will be able to do upon completion of the course. They establish a foundation upon which the rest of the course is based.

2.1: Objectives describe measurable outcomes
2.2: Module-level objectives are consistent with course-level objectives
2.3: Learning objectives are prominently located in the course
2.4: Relationship between learning objectives and activities are clearly stated
2.5: Learning objectives are suited to level of the course

Assessments

QM Standard 3 – Assessments are integral to the learning process; designed to evaluate learner progress in achieving the learning objectives.

3.1: Assessments measure the achievement of the learning objectives
3.2: Grading policy is stated clearly at the beginning of the course
3.3: Specific and descriptive criteria are provided for evaluation
3.4: Assessments are sequenced, varied, and suited to course level
3.5: Multiple opportunities to track learning progress with timely feedback

Instructional Materials

QM Standard 4 – Instructional materials enable learners to achieve stated learning objectives

4.1: The instructional materials contribute to the achievement of the stated learning objectives
4.2: The relationship between the use of instructional materials in the course and completing learning activities is clearly explained.
4.3: The relationship between the use of instructional materials in the course and completing learning activities is clearly explained.
4.4 The instructional materials represent up-to-date theory and practice in the discipline.
4.5 A variety of instructional materials is used in the course.

Nebraska
Lincoln

Learning Activities & Interactions

QM Standard 5 – Learning Activities and Learner Interaction:
Learning activities facilitate and support learner interaction and engagement.

5.1 The learning activities promote the achievement of the stated learning objectives or competencies.
5.2 Learning activities provide opportunities for interaction that support active learning.
5.3 The instructor's plan for interacting with learners during the course is clearly stated.
5.4 The requirements for learner interaction are clearly stated.

Nebraska
Lincoln

Course Technology

QM Standard 6 – Course Technology: Course technologies support learners' achievement of course objectives or competencies.

6.1 The tools used in the course support the learning objectives or competencies.
6.2 Course tools promote learner engagement and active learning.
6.3 A variety of technology is used in the course.
6.4 The course provides learners with information on protecting their data and privacy.

Nebraska
Lincoln

Design for Online Delivery

Quality Matters
• QM Rubric for evaluating course design

	QM Higher Education Rubric, Sixth Edition
General Standards	8
Specific Review Standards	42
Essential Specific Review Standards	23
Very Important Specific Review Standards	12
Important Specific Review Standards	7
Total Points Possible	100
Points Required to Meet Expectations (in addition to meeting all essential Specific Review Standards)	85

Design: Learning Activities and Materials

Discussion Board: Questions
• Northouse writes, "In any decision-making situation, ethical issues are either implicitly or explicitly involved." Give an example of a leadership decision that has explicit and implicit ethical dimensions.

• Over the past several weeks, we have covered many topics related to leadership. Please take some time to think about ideas, models, and discussions you have learned from, up to this point. List 2 or 3 specific models, ideas, concepts, theories, or practices you have learned in this course that challenged you, "blew your mind", or changed the way you think about leadership. Explain your answer. Have you practiced any of what you have learned? In what situation? Elaborate and discuss.

Design: Learning Activities and Materials

Discussion Board: Questions
• Contrast the experiences of in-group and out-group members. Have you observed groups that fit these descriptions in an organization (past or present) in which you work or with which you are affiliated? Give specific examples to support your observations.

• Defend or refute: Servant leadership should be conceptualized as a behavior rather than a trait. Justify your answer by citing the textbook, articles, and other sources that you have consulted.

Master of Engineering Management (MEM)

Background

• Fully Online
• 30-credit hour
• Non-thesis, professional
• 8-week classes (2 per semester)
• Partner with the Business College
• 72 students

Engineering Leadership – Fall B, 2017
8 weeks to get it ready
No good books on the market

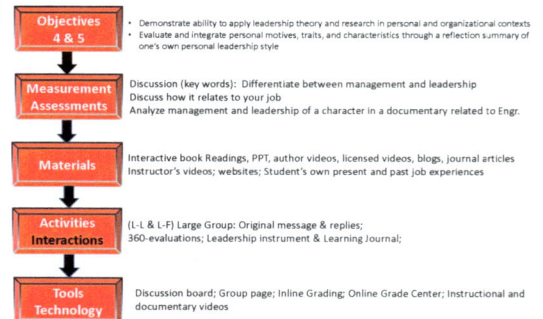

Objectives 4 & 5	• Demonstrate ability to apply leadership theory and research in personal and organizational contexts • Evaluate and integrate personal motives, traits, and characteristics through a reflection summary of one's own personal leadership style
Measurement Assessments	Discussion (key words): Differentiate between management and leadership Discuss how it relates to your job Analyze management and leadership of a character in a documentary related to Engr.
Materials	Interactive book Readings, PPT, author videos, licensed videos, blogs, journal articles Instructor's videos; websites; Student's own present and past job experiences
Activities Interactions	(L-L & L-F) Large Group: Original message & replies; 360-evaluations; Leadership instrument & Learning Journal;
Tools Technology	Discussion board; Group page; Inline Grading; Online Grade Center; Instructional and documentary videos

Learning Objectives and Outcomes

- Describe the fundamental differences between management and leadership, and how each concept affects organizational performance *(New York Underground)*
- Explain how personality traits, motives, and personal characteristics influence leader behavior *(Cases)*
- Articulate the components of several leadership theories including Transactional and Transformational leadership, Charismatic leadership, models of power and influence, Leader-Member Exchange theory, and Motivation theory *(Discussion)*
- Demonstrate ability to apply leadership theory and research in personal and organizational contexts *(Endurance)*
- Evaluate and integrate personal motives, traits, and characteristics through a reflection summary of one's own personal leadership style *(Personal Leadership Profile)*

QM Standards for Online & Blended Courses

8 General Standards (5 critical for alignment)
1. Course Overview & Introduction
2. Learning Objectives
3. Assessment and Measurement
4. Instructional Materials
5. Learning Activities & Learner Interactions
6. Course Technology
7. Learner Support
8. Accessibility and Usability

Learning Materials

Course Capstone Project

Endurance – Case and Documentary:
Review of all leadership styles studied in class
    Sir Ernest Shackleton – 1914 expedition (England to Antarctica)
    Evaluation of leadership style of Shackleton and main characters

- Kohen, Nancy. (2010). *Leadership in Crisis: Ernest Shackleton and the Epic Voyage of the Endurance*, Harvard Business School No. 9-803-127. Boston, Harvard Business school Publishing.
- https://www.youtube.com/watch?v=oyQRHHHXntc

Learning Materials

Course Capstone Project

Personal Leadership Profile: **Synthesis** & **Integration** (Video project)
- Book and articles
- Leadership instruments
- Learning journals
- 360 evaluations

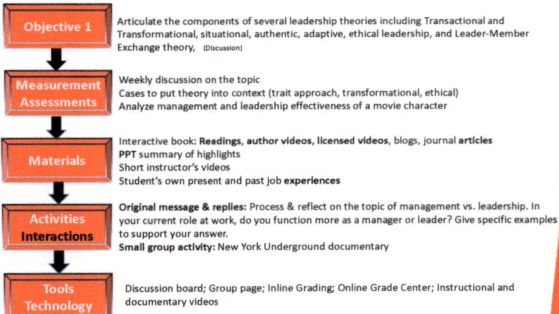

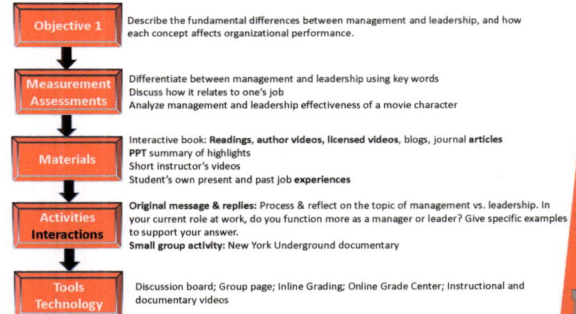

Objective 1	Articulate the components of several leadership theories including Transactional and Transformational, situational, authentic, adaptive, ethical leadership, and Leader-Member Exchange theory, (Discussion)
Measurement Assessments	Weekly discussion on the topic. Cases to put theory into context (trait approach, transformational, ethical) Analyze management and leadership effectiveness of a movie character
Materials	Interactive book: **Readings, author videos, licensed videos**, blogs, journal **articles** PPT summary of highlights Short instructor's videos Student's own present and past job **experiences**
Activities Interactions	**Original message & replies:** Process & reflect on the topic of management vs. leadership. In your current role at work, do you function more as a manager or leader? Give specific examples to support your answer. **Small group activity:** New York Underground documentary
Tools Technology	Discussion board; Group page; Inline Grading; Online Grade Center; Instructional and documentary videos

---

Objective 1	Describe the fundamental differences between management and leadership, and how each concept affects organizational performance.
Measurement Assessments	Differentiate between management and leadership using key words Discuss how it relates to one's job Analyze management and leadership effectiveness of a movie character
Materials	Interactive book: **Readings, author videos, licensed videos**, blogs, journal **articles** PPT summary of highlights Short instructor's videos Student's own present and past job **experiences**
Activities Interactions	**Original message & replies:** Process & reflect on the topic of management vs. leadership. In your current role at work, do you function more as a manager or leader? Give specific examples to support your answer. **Small group activity:** New York Underground documentary
Tools Technology	Discussion board; Group page; Inline Grading; Online Grade Center; Instructional and documentary videos

Feedback is a Gift

Marcia Dority Baker (UNCA)
Casey Nugent (UNCA)

In August 2017, the AT (Academic Technology) Advisory Committee was created. This group was initiated as an opportunity for faculty input on academic technologies for teaching and learning at the University of Nebraska-Lincoln. The advisory committee provides feedback on policy considerations and strategic initiatives. Advisory committee members include faculty from each college who meet with the AT staff twice a semester. There is currently one subgroup assisting with data analytics and learning outcomes via the data from Canvas (LMS at the University of Nebraska). The AT Advisory Committee has been beneficial to Academic Technologies for assistance in policy for the Digital Learning Commons (DLC), feedback on the learning analytics dashboards for faculty and input on services such as Student Response Systems (SRS). This presentation will discuss the process for starting the AT Advisory Committee and projects and/or initiatives the committee is working on.

This presentation featured:

- the relationship between faculty and academic technologies.
- initiatives currently underway that allow instructors to fully utilize available technology within the classroom.
- how to access Canvas Analytics and other data tools.

Feedback is a Gift
▸▸▸
Marcia L. Dority Baker
Casey Nugent

Nebraska

Purpose – AT Advisory Committe
▸▸▸
- Created in August 2017
- Opportunity for faculty input on academic technologies for teaching and learning.
- Provide feedback on policy considerations and strategic initiatives.
- Advisory committee members include faculty from each college
- Meet with the AT staff twice a semester.

Nebraska

Collaboration

Academic Technologies - UNL

- DLC
- Learning Spaces/Classrooms
- Teaching Technologies
- Learning Analytics (from all three)

ONE UNIVERSITY. FOUR CAMPUSES. ONE NEBRASKA.   Nebraska

Initiatives

- Digital Learning Center policy
- Learning Spaces
- OER and Kelly Grant
- Pilot opportunities with AT services
- Canvas Learning Analytics and Data

ONE UNIVERSITY. FOUR CAMPUSES. ONE NEBRASKA.   Nebraska

Digital Learning Center

- Aligning DLC hours with Final exam schedule
- Setting foundation for successful testing experience in DLC
- Scanning Services assisting with Course Evaluation Taskforce – EvaluationKit
- Automating Scanning Services requests

Learning Spaces & Classrooms

- Collaborative classrooms
- Summer updates
- Enhanced tech for VidGrid and Zoom services
- Student response systems

Emerging Technologies

- OER opportunities
- Kelly Grant
- STAR initiative

Successful Teaching & Affordable Resources N

Learning Analytics

- Objective: Improve outcomes for students
  - Provide feedback to students and instructors
  - Work with instructors to provide information to assist advisors
  - Collaborate with student success initiatives and research

Learning Analytics Focus Group

- Agile, exploratory philosophy
- No monolithic approaches
- Connecting, discovering, sharing, evaluating, and strategizing

Learning Analytics Priorities

- Integrating Data Sources
- Discovering Successful Learning Patterns
- Disseminating Information
  - Dashboards
  - Discussions

Integrating Data Sources

- Digital Learning Commons Exam Reservation System
- HuskerScan
- ACE Outcomes
- Unizin Data Platform

Integrating Data Sources

- Unizin
  - Member since June 2015
  - Consortium of 13 universities and university systems
  - Collaboration and interaction
  - Substantial savings on Canvas, TurnItIn, and others

Integrating Data Sources

- Unizin Data Platform
  - Flexible and efficient data integration
  - Anonymized data from other participating schools
    - Ohio State, Penn State, University of Michigan, University of Iowa, etc.
    - Not only helpful for benchmarking, but also for model development

Integrating Data Sources

- Leveraging the advisory and focus groups
  - Feedback
  - Direction & Prioritization
  - Communication

Successful Learning Patterns

- Example: Student Engagement Models
  - Discussion Participation
  - Course Level LMS Interactions
- Feedback
  - Distinction between differing discussion board uses
  - Instructor-specified participation cutoffs

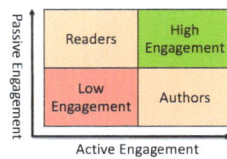

Successful Learning Patterns

- Other research
  - Student Learning Process
  - Canvas Course Design Efficiency
  - Course Level Navigation
  - ACE Outcomes

Disseminating Information

- Engagement and Discussion
  - AT Advisory and Learning Analytics groups
  - Instructional Design group
  - Ad hoc campus engagements
- Reports and Dashboards
  - Canvas Analytics
  - Tableau

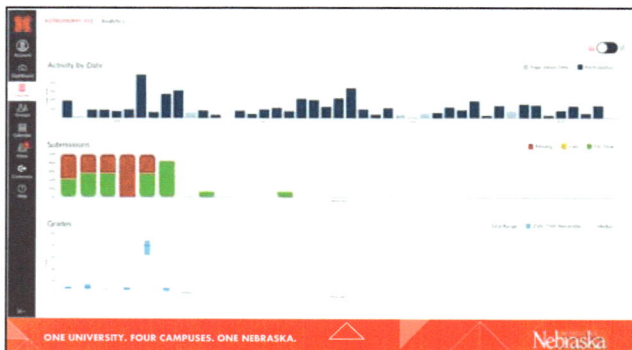

Canvas Analytics

Activity by Date Pageviews and participation

Canvas Analytics

Submissions: Missing, late, and on-time

Canvas Analytics

Grades: Box and whisker for each assignment

Tableau

What is Tableau, anyway?
- Flexible & Interactive Data Visualization

Why are we using it?
- Versatile intake of data sources
- Highly usable interface
- Many successful implementations in Higher Ed

Goal: Customizable Canvas-Embedded Interactive Reports

Example: LTI Activation

Summary

Feedback is a Gift
- Policy review
- Strategic direction
- Task-level assistance
- Bi-directional communication
- Connecting interests

FACULTY FEEDBACK:
"SERVING ON THIS COMMITTEE HELPS ME FEEL LIKE PART OF A LARGER TEAM WORKING TOGETHER TO SUPPORT STUDENTS AND MAKE GOOD THINGS HAPPEN."

THANK YOU!

Advancing Technology in Education at the University of Nebraska

Student-Centered Blended Learning: The HyFlex Approach to Blended Learning

Benjamin R. Malczyk, Ph.D. (UNK)
Dawn Mollenkopf, Ph.D. (UNK)

While blended learning provides some flexibility to students, the implementation of blending learning is generally faculty or instruction driven. The HyFlex blended learning model is an alternative approach to blended learning that places power in students' hands. This presentation will outline several approaches utilized to provide more flexible approaches to learning for students. HyFlex blended learning is one such approach. The HyFlex blended approach allows each individual student to decide their modality of instruction (face-to-face, synchronous online video-conference or asynchronous coursework) on a week-to-week basis. This allows students to make decisions based on their needs and context for each class period. The presentation will summarize student choices and preferences regarding the modality of instruction as well as the key benefits and challenges of a student-centered approach to blended instruction. Students participating in a HyFlex blended learning experiment were able to attend in-class or online in a way that matched their needs for each given week.

This presentation featured:

- the movement in higher education toward student-centered approaches to instruction.
- various approaches to blended instruction and define HyFlex blended learning.
- the benefits and challenges of utilizing student-centered approaches to blended learning, including the HyFlex blended learning tool.

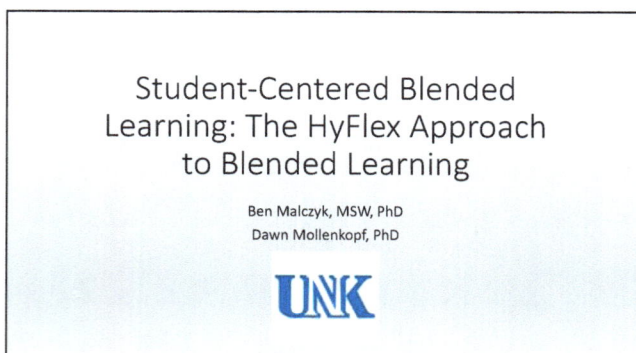

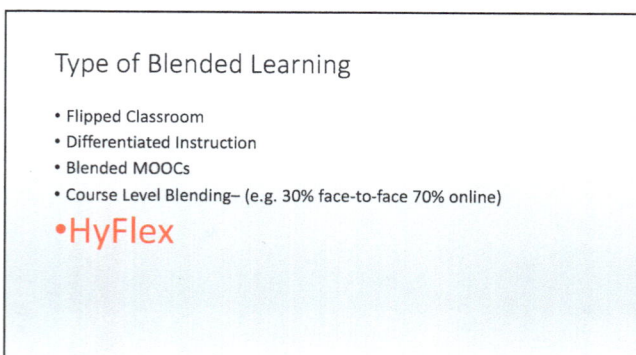

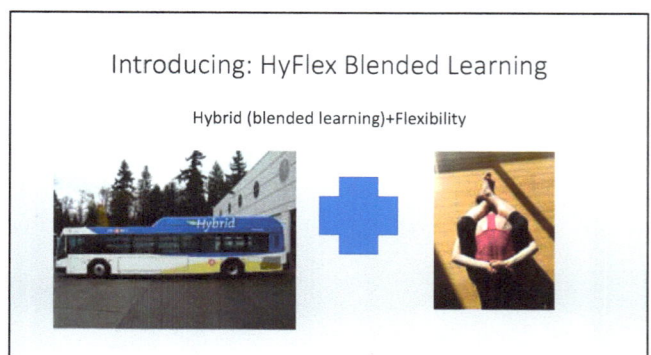

Students make a choice for *EACH* class session

Face-to-face

Asynchronous Online

Video Conference

LEARNING

Specific example—5 week Pilot

- Pre and post-test to assess student choice of modality
- Lesson plan for 1 week
  - Module on policy performance evaluation
    - Quiz
    - Worksheet or LMS
    - Small group discussions/individual tasks
      - E.g.– watch a brief video comparing US and Korean healthcare systems– which would you prefer, why?

Student Rankings of Preferred Modalities

Modality	If students could only choose 1 modality	Pretest average rank	Posttest average rank
Face-to-face	10	1.3	1.5
Video conference	1	2.6	2.9
Asynchronous	7	2.1	1.6

Student Predicted Attendance Compared to Actual Attendance

	Week 1		Week 2		Week 3		Week 4		Week 5	
	Predicted	Actual	Predicted	Actual	Predicted	Actual	Predicted	Actual	Predicted	Actual
Face-to-face	11	8	12	6	10	7	8	6	12	6
Online	5	10	4	12	6	11	9	12	5	12
Video Conference	2	0	2	0	2	0	1	0	1	0

Student Blending Choices

Ratio of Online to Face-to-face	Number of Students	Percent of Students
100% Online 0% Face-to-face	8	44%
80% Online 20% Face-to-face	1	6%
60% Online 40% Face-to-face	4	22%
40% Online 60% Face-to-face	0	0%
20% Online 80% Face-to-face	2	11%
0% Online 100% Face-to-face	3	17%

Exploring HyFlex Learning

- Senior level class: Blended qualities in class already
  - Some assignments online
  - Field-based so some days students already in field during class time
- First two weeks—required class meetings; then took vote weekly to meet or not meet as a class
- Students also voted on topics to discuss (e.g. assignments, readings, field experience needs)
- Voted for a culminating event

Impact of Student-Centered Blending

- Students stayed engaged online and face to face—performed well on assignments/tasks and in field
- Students took charge of their learning--accountability
  - Created their own social network—communicating with peers on their own for class assignments and events
  - Introduced topics not part of class but pertinent to their careers
  - Spontaneously taught peers in class

Take Away Messages

- Hyflex can add an additional measure of student engaged learning to a class
  - Focused class time on what students needed
  - Learning beyond assignments
- Challenges:
  - online and face to face portions have to be equally well prepared
  - Professor must be open to multiple topics—relinquish control

Summary

- Pros
  - Students get what they need;
  - May work best for students;
  - Built in accommodations (UDL) in recognition of life's complexities
  - Same outcomes but more student satisfaction (Miller, Risser, & Griffiths)
- Cons
  - Time to plan—somewhat like teaching 2 courses; planning and preparation; technical ability; may have to be flexible with learning activities; may not work for all classes– for example, in social work we teach courses on clinical skills

Enhanced Online Student Engagement & Learning through 'Video Theater'

David Harwood, Ph.D. (UNL)

Learn about an online instructional approach to increase student engagement and promote asynchronous teamwork, which allows for self-exploration on topics of interest while facilitating passive but motivated exposure to content. The 'Video Theater' program incorporates student-self short videos that allow students to propose, review, nominate and vote for top-videos across a range of themes in an online course. Students work individually, in teams and within the full class to identify and filter through a range of content-rich videos from YouTube, National Geographic, NASA, History Channel, etc. These top videos are brought to the front and share with online classmates. This program affords students the flexibility to follow their interests, share visually rich information with peers and explore new content passively and effortlessly into their personal learning sphere and that of their team members.

This presentation featured:

- an effective approach to foster student engagement through video.
- the effective selection of video themes that would focus student searches within the course content structure.
- how to slightly adapt course content in order to introduce related content and provide new video content that can be incorporated into future online delivery.
- how students, working individually and in teams, progressively advance top videos for full-class viewing and commentary.

"I often benefitted from searching for videos. I often got 'sucked in' and found myself watching lots of videos because there are so many interesting ones."

"I really enjoyed learning about Antarctica in a visual way. It allowed us to search topics we are personally interested in, while also learning about the theme."

"The opportunity to search for my own videos made me engage and look for some unique info."

"Cool to look into the subjects we were learning about, but to explore further with outside video sources."

"I absolutely loved dedicating a class period to learning more about Antarctica in a modern way. This reinforced our use of technology in a proficient way."

MONDAY ANTARCTIC THEATER

Present outstanding videos that were located, reviewed, selected by students.

Four different themes on a 2-week rotation, 1 week break between.

Themes track course content, and progressively elevate in complexity.

WINNING videos in 4 categories are shown during class Monday.

Move to THEATER or Lecture Hall to benefit from Large SCREEN & Big SOUND!

Video length target of <5 minutes, but not to exceed 7 minutes.

CANVAS 'Discussion' work-space to PROPOSE videos, comment on videos proposed by team-mates, and NOMINATE the Team's 'Final Selections'.

Instructor's easy tracking and engagement with the discussion.

CANVAS preserves a history of discussions and selection process.

The process to select videos: working in teams of 3 to 4 students

Students **SEARCH** the internet (watching a minimum of 8 videos) to find four (4) high-quality videos, and **PROPOSE** to their team. These can be in in any of the 4 categories listed below (but they must 'propose' in at least 2 of the categories.

All Team members will view all proposed videos, **DISCUSS** (on CANVAS); team will agree on the best 5 videos to **NOMINATE** into a larger pool that includes top-ranked videos from 4 other teams.

The nominated videos will be reviewed and rated by half of the class (the 'left-side' vs. 'right-side' of the classroom). Students will **VOTE** to select the best video in each of the 4 categories to help select the "FINALISTS" to be viewed during ANTARCTIC THEATER.

For each of the four Themes, students watch ~50 videos!!!

SEARCH for videos 8 (min.) **to 12** (average)	**12**
REVIEW team members' proposed videos (numbers are for teams of 4 students)	**12**
VOTE on ½ class video nominations (numbers are for 5 teams)	**20**
ANTARCTIC THEATER in-class viewing	**12**
	56 TOTAL

>200 videos watched for all 4 themes!!

Theme A: 'Antarctic Scenery' – and 'cool' environments

Categories:

(A1) where rock meet ice - exposed mountains and valleys with ice and glaciers;

(A2) where ocean meets air - Southern Ocean storms and waves, fury of intense energy;

(A3) where floating ice meets the ocean - sea-ice and ice shelves, icebergs;

(A4) general scenery themes - not covered above.

Bringing Antarctica into the Classroom
"Sense of Place"

Theme B: 'Humans on ice'

Categories:

(B1) Scientists in the field, at a base, on a research ship, etc.;

(B2) Tourists in Antarctica; polar adventurers; etc.;

(B3) Extreme conditions: 'lucky' to be alive (seeing cool stuff, doing stupid stuff, etc.);

(B4) Explorers of the past – historical journeys.

Seeing themselves there

Theme C: 'Doing Science' – science & technology in Antarctica

Categories:

(C1) visualizing and remote sensing of what exists within and beneath the ice;

(C2) instruments, tools and projects to sample, analyze and study ice, rocks & sediments;

(C3) tours of research facilities (ships, stations, laboratories, etc.);

(C4) getting there: transportation (water, air, land, snow).

"Tools of the Discipline"   Exploration in the Modern Era

Theme D: 'Signs of Change' – past records & future projections

Categories:

(D1) our warming world - videos presenting changes in ocean, air, land, and ice;

(D2) melting ice & warming seas – focus on changes in Antarctic ice and sea level rise;

(D3) impactful and thought-provoking alerts regarding present and future changes;

(D4) international/national discussions & actions, or predictions by numerical models.

"Taking Ownership"   Tracking the Past - Perceiving the future

---

Discussions section provides space for:
PROPOSALS to team (individual)
REVIEW & DISCUSSION (team)
NOMINATIONS to ½ class (team)

Google FORMS enables easy management of:
VOTING Sheet and url links to videos
Records and Reports student VOTES

You Tube PLAYLIST utility in organizing:
FINAL VIDEOS list and order
THEATER PROGRAM list with urls
SHARING, ANNOTATING & ARCHIVING videos

---

canvas → Google FORMS → You Tube PLAYLIST

Let's take a TOUR through the process of developing

ANTARCTIC
VIDEO
THEATER

TEMPLATE for: Individual's Video *PROPOSALS*
to 6 entries into CANVAS Discussion)

Proposed VIDEO # ___

1. Theme and Category (e.g. A2)
2. Video URL or web-link:
3. Video Title or Brief Description (in BOLD font); if it has a title, don't rename it.
4. Time – duration in minutes;
5. Justification for proposing this video;
6. Your Name and Team number

TEMPLATE for: Team *NOMINATION* of 5 videos:

Nominated Video # ____ ; Team ____

1. Theme and Category
2. Video URL or web link:
3. Video Title or Brief Description (in BOLD font); if it has a title, don't rename it.
4. Time – duration in minutes;
5. Team member(s) who first PROPOSED video

Here is an example for you to follow. Copy the template for a PROPOSAL from the text above and paste it in the REPLY box to start.

**PROPOSALS
&
NOMINATIONS**

Proposed VIDEO #1

1. Theme and Category (e.g. A3) "where floating ice meets the ocean"
2. Video URL or web-link: https://www.youtube.com/watch?v=MPFj2t7Fo24
   An Endless Expanse of Pancake Ice.
3. Video Title
   An Endless Expanse of Pancake Ice
4. Time - duration in minutes: 1:27
5. Justification for proposing this video: This video shows the rolling nature of a frozen ocean and pancake ice.
6. Your Name and Team number: David Harwood, Team 0

Unread

↩ Reply

PROVIDING A TEMPLATE

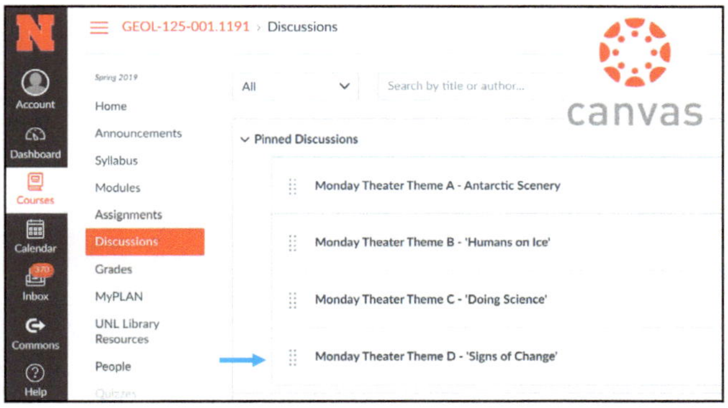

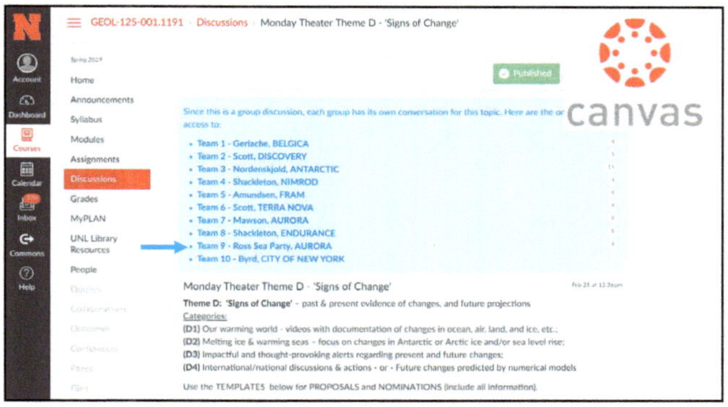

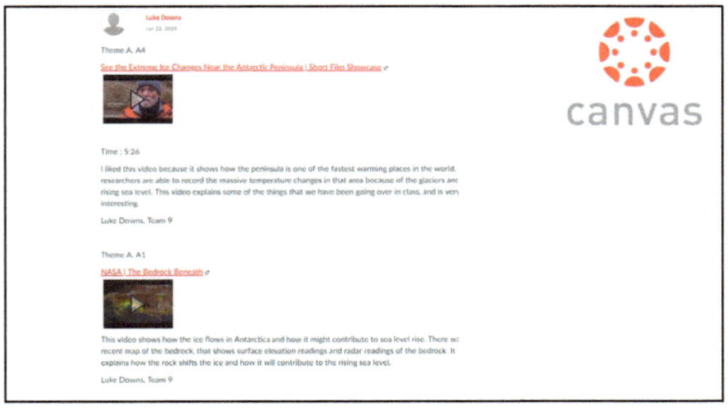

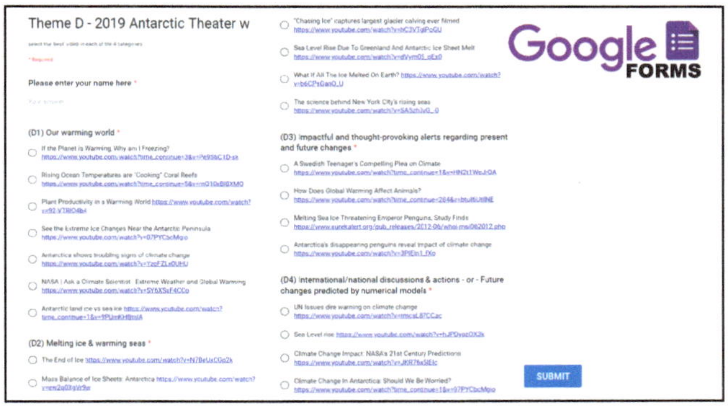

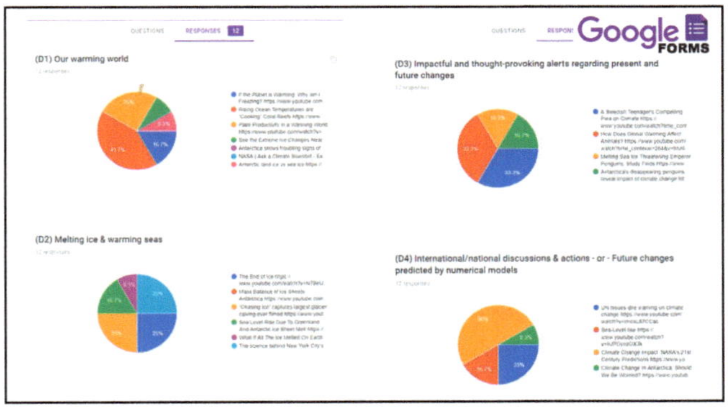

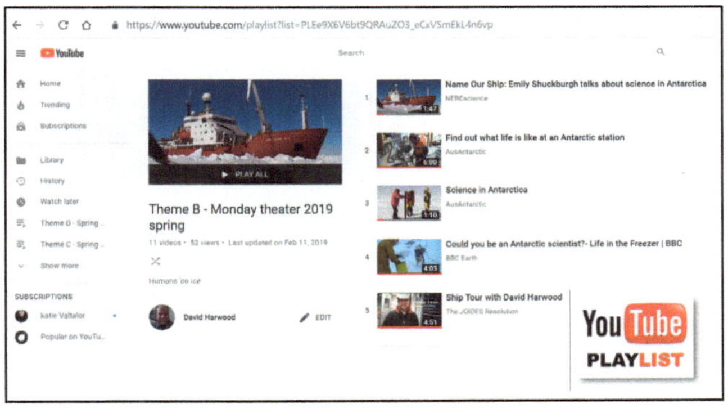

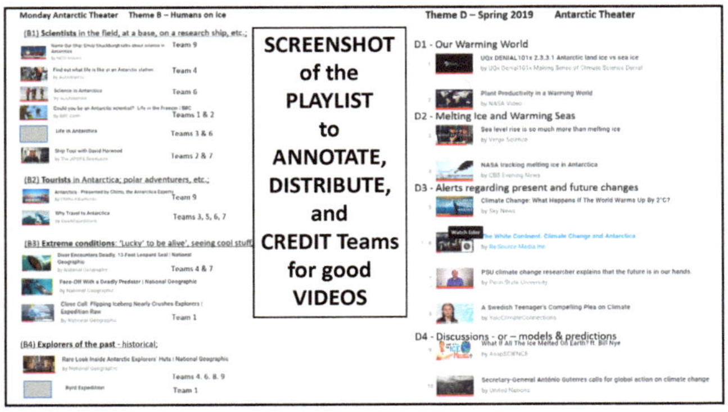

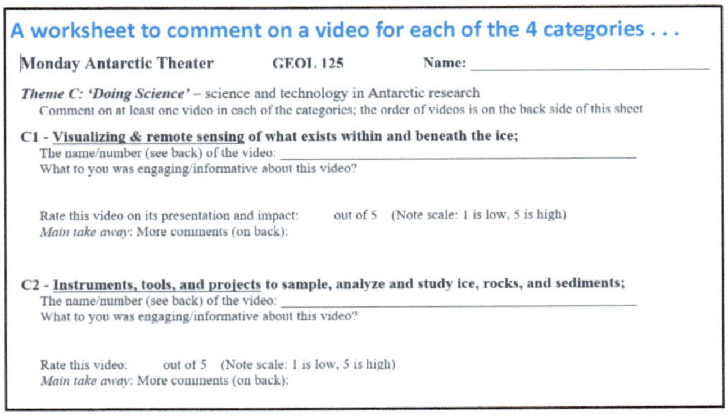

SCREENSHOT
of the
PLAYLIST
to
ANNOTATE,
DISTRIBUTE,
and
CREDIT Teams
for good
VIDEOS

A worksheet to comment on a video for each of the 4 categories . . .

Monday Antarctic Theater   **GEOL 125**   Name: _____

Theme C: 'Doing Science' – science and technology in Antarctic research
Comment on at least one video in each of the categories; the order of videos is on the back side of this sheet

C1 - Visualizing & remote sensing of what exists within and beneath the ice;
The name/number (see back) of the video: _____
What to you was engaging/informative about this video?

Rate this video on its presentation and impact: out of 5   (Note scale: 1 is low, 5 is high)
Main take away: More comments (on back):

C2 - Instruments, tools, and projects to sample, analyze and study ice, rocks, and sediments;
The name/number (see back) of the video: _____
What to you was engaging/informative about this video?

Rate this video: out of 5   (Note scale: 1 is low, 5 is high)
Main take away: More comments (on back):

. . . concludes with questions derived from a 1st day pre-test

General comments:

Rate the level of agreement with the following statements: (rank between **1** (disagree fully) to **5** (agree fully))

	1	2	3	4	5
Research in Antarctica provides important information to society	1	2	3	4	5
It is amazing what scientists interpret about Antarctica's processes & history	1	2	3	4	5
National support for Antarctic research is important and should continue	1	2	3	4	5
Scientists are motivated by curiosity more than fame & future grant funding	1	2	3	4	5
Technological advances have enhanced perceptions of our dynamic planet	1	2	3	4	5

From your viewing of all videos for Theme C - describe an element of Antarctic research (vehicle, instrument, tool) that seems "out-of-this-world" & future sci-fi: or describe what surprised you about Antarctic facilities and tools:

You have 5 minutes with a member of Congress to tell them about Antarctic research and why the US should continue to fund research and a presence there, what would be the main points of your message?

Your students will 'mine' the internet for good, interesting videos;

Your students will explore their own interests;

Your students will locate exceptional videos for your future use.

"I thought it was a nice break from regular class, while still learning."

"I really liked Antarctic Theater and the way it makes you go out and search for educational videos, and learn things that interest YOU."

Thank you for your attention!
I'd welcome your questions and comments.
I hope something like ANTARCTIC VIDEO THEATER might work for your courses

Dr. David M. Harwood dharwood1@unl.edu
Professor, and TM and EE Stout Chair of Stratigraphy
Dean's Fellow of the College of Arts & Sciences Teaching Academy
Director, Antarctic Science Management Office

Department of Earth & Atmospheric Sciences
University of Nebraska-Lincoln
Lincoln, NE 68588-0340 USA

Taking Public Speaking Classrooms Up a Notch with Digital Video Recording

Rick Murch-Shafer (UNO)

Technology in the classroom can greatly enhance instruction for most disciplines. But as technology gets outdated, how do you manage the equipment and end users to maintain or improve upon a tried and true classroom environment? UNO is currently replacing aging technology in Public Speaking classrooms with modern digital video recording technologies. Come and learn about the design and approach UNO is using to maintain a consistent process and usher in new, more sustainable technologies to take it into the future.

This presentation featured:
- classroom video capture technologies.
- the lesser known features of VidGrid.
- tips on how to deal with fear of change within the context of updated/new technology.

Featured Extended Presentation

Emerging Technology Trends:
Virtual Reality and Artificial Intelligence

Featured Speaker: Bryan Alexander

This session will focus on the two major technological forces likely to impact online education in this extended session. We will explore the possibilities of virtual reality (VR), examining its current pedagogical aspects and institutional support structures. We will also examine upcoming virtual reality uses, including social VR as well as VR as an alternative to videoconferencing. Secondly, we will address artificial intelligence (AI) by reflecting on its current and emerging uses in higher education, from chatbots to AI-powered teaching software and student data analytics. We will continue on a prospective note, looking at trendlines and the history of technology to anticipate medium-term uses of AI in academia. Our conclusion steps back to consider how online education could change as AI starts transforming the labor market and society as a whole.

This presentation featured:

- the impact of VR and AI on higher education.
- the pedagogical aspects of VR.
- how AI-powered technology can impact the student experience.

Online Program Lead Nurturing Panel

Bob Mathiasen, Ph.D. (UNL)

Stacey Schwartz (UNK)

Angie Tucker (UNMC)

Alex Zatizabal Boryca (UNCA)

In 2018, an Online Program Lead Nurturing Specialist was hired for each campus. The focus of these positions is to engage in proactive outreach to prospective online students with the goal of helping them through the application and enrollment process so more students enroll in online programs at NU.

This presentation featured:

- the Lead Nurturing Specialist positions.
- the impact these positions made to student enrollment.
- how these positions complement their admission and departmental recruitment efforts.

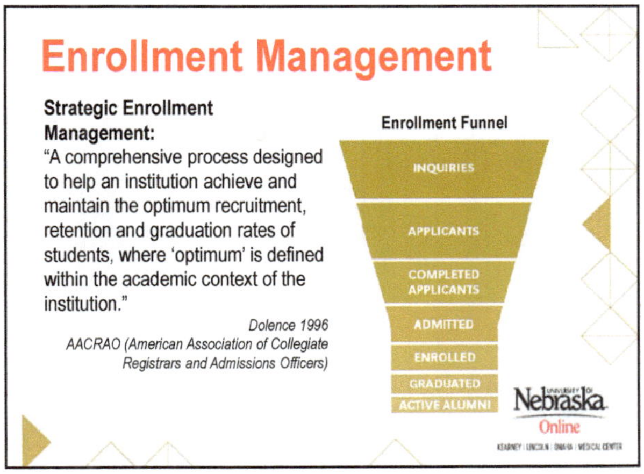

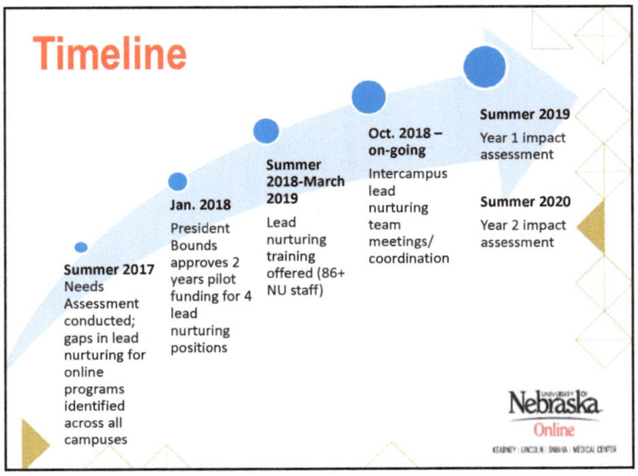

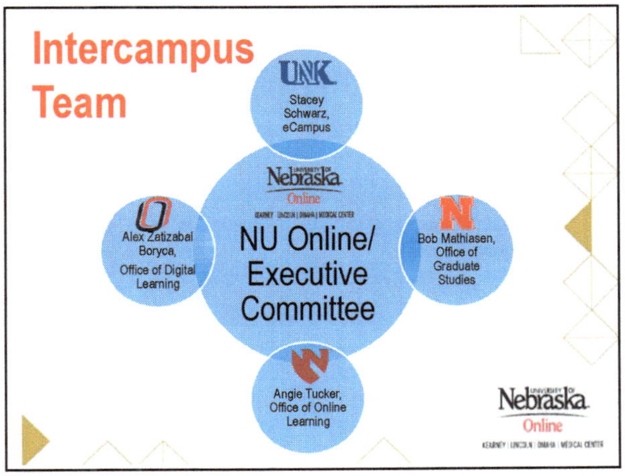

Plan, Enroll, Progress: Integrated Planning & Advising for Student Success

Steve Booton (UNL)

Bill Watts (UNL)

The lack of proactive academic planning tools and supporting integration between systems will not facilitate our goal of significantly increasing our retention and graduation rates. Students need to see their whole academic plan, know when they are off-track and develop ownership of their own curricular experience. In support of this idea, faculty and staff plan capacity for course demand, track student academic progress and provide real-time student intervention. To achieve these objectives, a variety of tools are being developed, purchased and integrated through a single portal that empowers students to take responsibility for their choices while encouraging departments to use newly available data to meet planned student demand.

This presentation featured:

- empowering students by allowing access to their whole academic plan.
- data that helps students become more proactive about their planning.
- new tools and systems incorporated in a student planning system.

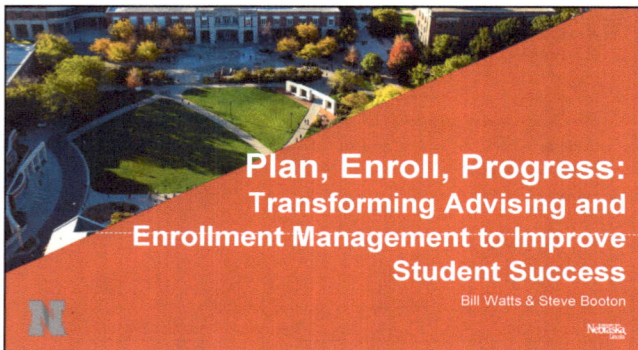

Plan, Enroll, Progress:
Transforming Advising and Enrollment Management to Improve Student Success

Bill Watts & Steve Booton

Challenge

The lack of proactive academic planning tools and minimal integration between student information systems will not facilitate our goal of significantly increasing our retention and graduation rates.

Students Need To

- See the "whole" picture for their academic plans
- Know when they are off track
- Receive early guidance for "Plan B" options
- Develop accountability and ownership of their own curricular experience

Faculty & Staff Need

- Course demand data to plan for capacity
- Automated systems to identify off track students
- Ongoing and tailored student interventions
- Displays that provide real time data on their students

Our Plan

Pursue a project to simplify the complexity of course and student planning systems to enable the following:

- **Empower** students to take responsibility for choices in their best interest with the confidence their needs and goals will be met.

- **Encourage** departments to use newly available data to anticipate and meet planned student demand.

- **Promote** early intervention and a well articulated, campus-wide focus on student success.

System Modules

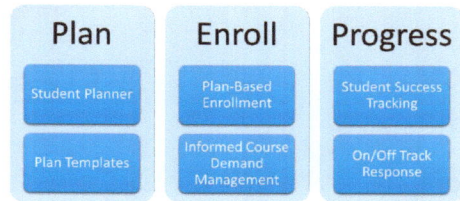

Plan	Enroll	Progress
Student Planner	Plan-Based Enrollment	Student Success Tracking
Plan Templates	Informed Course Demand Management	On/Off Track Response

Sample Degree Plans
Student selects a map by academic program

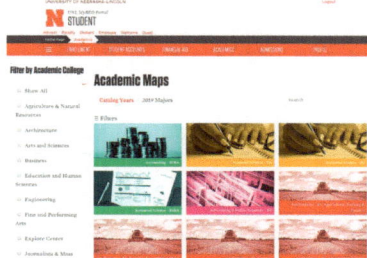

Sample Degree Plans
Student sees the typical 4-year plan

Degree Planner
Student moves courses into plan by term

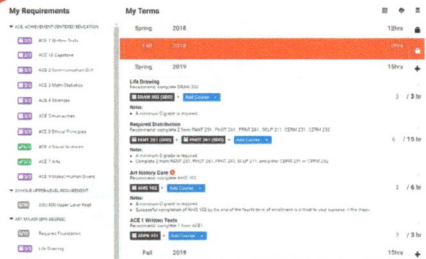

Plans in Degree Audit
Student planned courses updates degree audit

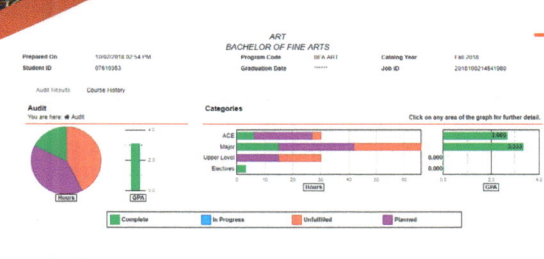

Enrollment Scheduler

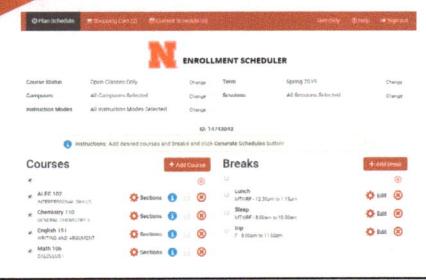

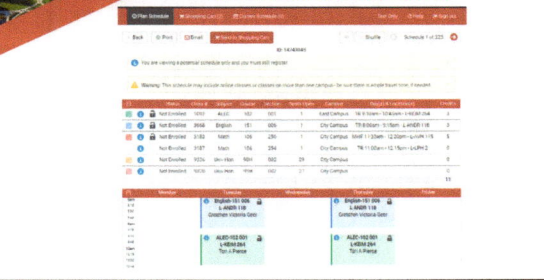

Enrollment Scheduler

Plan, Enroll, Progress & Student Status

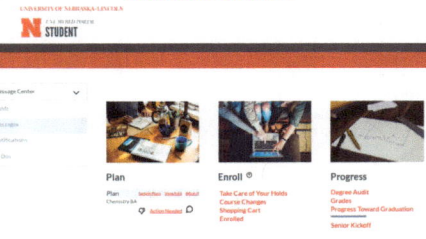

Next Steps

- Fall 2019 New Student Cohort
- Create course demand response system
- Create student on/off track response system
- Develop dashboards for advisors and program staff

360 Degrees of Geography

Nate Eidem, Ph.D. (UNK)

Steve McGahan (UNK)

Photography has been a standard aspect of physical geography, but as technology advances, it presents opportunities to increase learning in online, blended and face-to-face courses by incorporating more interactive visual imagery and video. These media can be invaluable in allowing students to see different facets of the physical world. Incorporating a virtual field trip component in an online course can be challenging, but methods of bringing students to new locations virtually are within the reach of faculty. The presenters will discuss the collection and use of 360 degree video and other forms of imagery to develop virtual field assessments for modules in an online physical geography course. The use of 360 degree cameras, drones and standard photography will be discussed, as well as how to find these types of assets without having to directly collect them.

This presentation featured:

- how the barriers to the integration and use of 360 degree video, imagery and other visual elements are shrinking.
- how the use of 360 degree video, imagery and other visual elements will become expected by the upcoming student population.
- how the integration of 360 degree video and imagery can give students a more engaging experience.

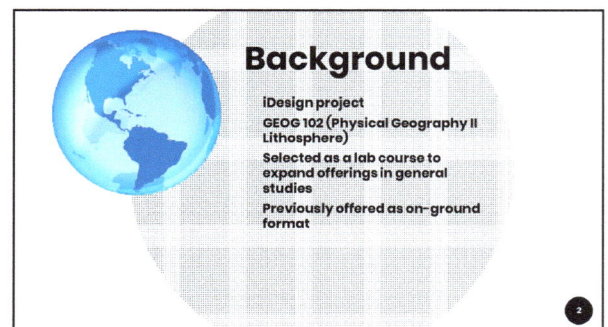

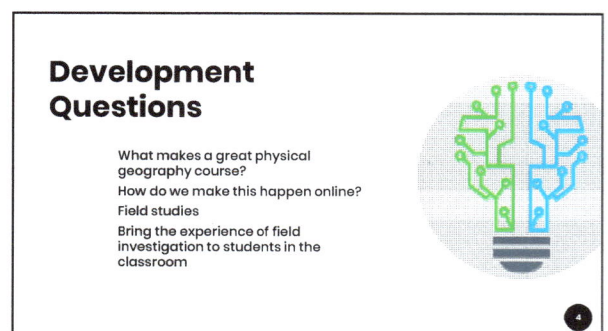

Primary Development Goals

- Increase student/content interaction
- Present content in a more engaging format
- Integrate kinesthetic and visual learning content
- Decrease costs for students
- Increase availability of general studies courses for degree students

5

2

Field Kit

Collection Equipment List

360° Video/Images

- GoPro Fusion
- Spare battery
- Micro SD cards
- Various grips
- Suction cup
- Remote control

7

Aerial Video

- DJI Mavic Air Combo
- 4 batteries (30 min flight time each)
- Replacement propeller blades
- Propeller guard
- Micro SD cards
- iPhone 6/iPhone 6s Plus
- Keep flying app up to date

8

Other Equipment

- GoPro Hero 2 (suction cup drives)
- Nikon D3400 (stills/video)
- Canon Power Shot (stills/video)
- Dell XPS Laptop (editing)
- External Hard Drive (backup)
- Compressed air
- Micro-fiber cloths
- Charging equipment

9

3

Integration of Video and VR

Content redefined

Video Integration

- Use of field video to enhance content and increase student/content interaction
- Virtual field studies
- Data collection points for geographic calculation assignments
- Research project

11

Virtual Reality Development

- Use of 360° camera to create investigation points for the course
- Integration of content in to the images/videos
- Create an immersive environment
- University of Maryland study on VR retention

12

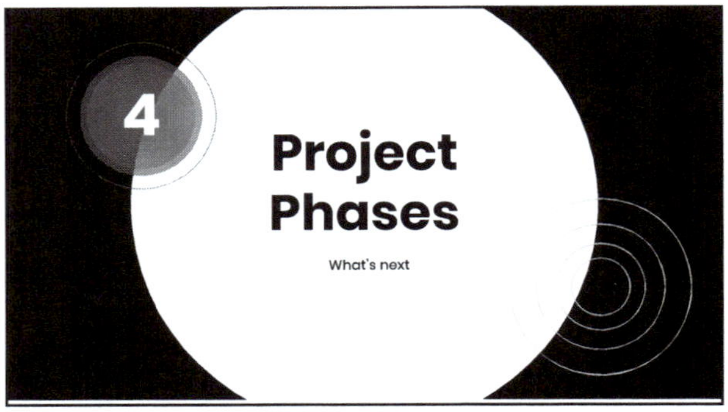

Project Phases

4

What's next

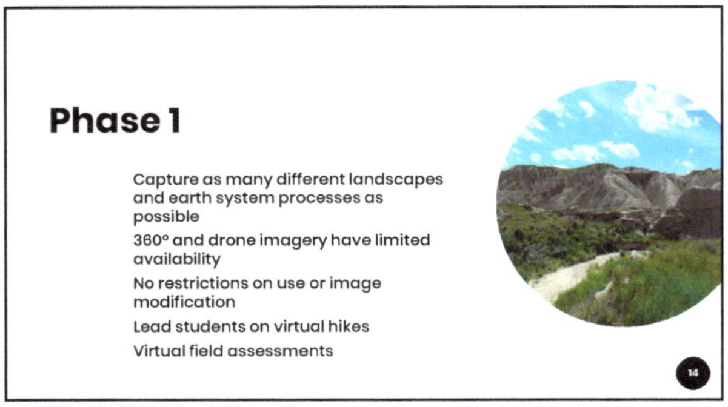

Phase 1

Capture as many different landscapes and earth system processes as possible

360° and drone imagery have limited availability

No restrictions on use or image modification

Lead students on virtual hikes

Virtual field assessments

14

Phase 2

Augmentation and storytelling
Stand alone edu-videos
Video quizzes
VR experiences

15

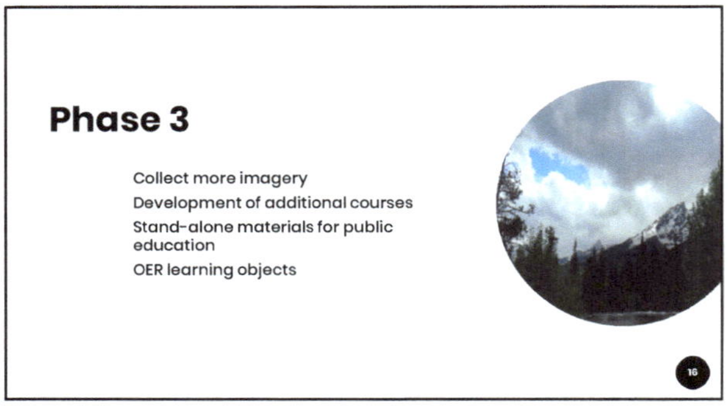

Phase 3

Collect more imagery
Development of additional courses
Stand-alone materials for public education
OER learning objects

16

Bringing Offline Geography to the Online Student

17

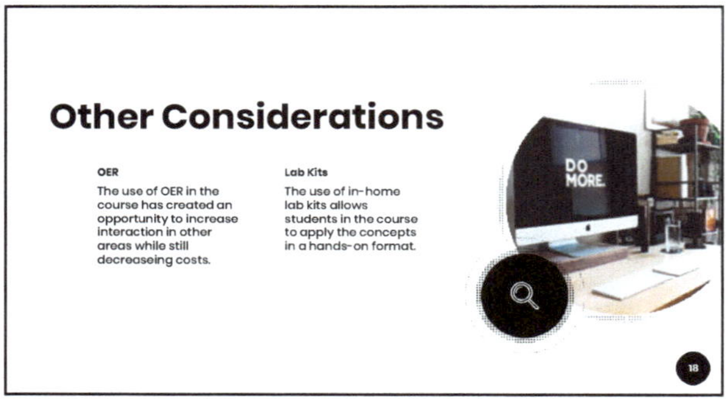

Other Considerations

OER
The use of OER in the course has created an opportunity to increase interaction in other areas while still decreaseing costs.

Lab Kits
The use of in-home lab kits allows students in the course to apply the concepts in a hands-on format.

18

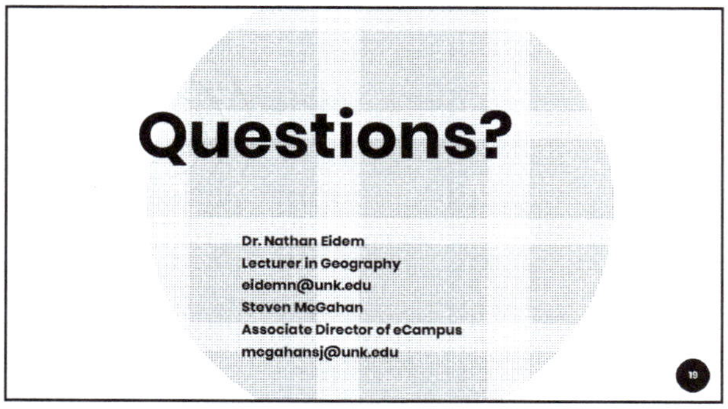

Questions?

Dr. Nathan Eidem
Lecturer in Geography
eidemn@unk.edu
Steven McGahan
Associate Director of eCampus
mcgahansj@unk.edu

19

Using Zoom to Reach a National Audience

Saundra Wever Frerichs, Ph.D. (UNL)

Outreach is important to many departments, but it is the focus and mission of Nebraska Extension. This focus on reaching audiences outside the University community shapes how we use technology and now others have the opportunity to learn from our experiences. Online outreach provides opportunities to attract highly engaged students to an academic program, to share research and broaden its impact and to provide value and service to all Nebraskans. This session will help participants broaden the audience for their work and increase their impact.

This presentation featured:

- potential audiences to reach via technology.
- effective practices that align with your goals for outreach and extension.
- strategies that can be implemented right away.

Internet Access in Nebraska

Loup county
2.9%

Lancaster county
97.3%

Broadband Service in Nebraska from Broadband Now
https://broadbandnow.com/Nebraska

CLICK 2 SCIENCE
click2sciencepd.org

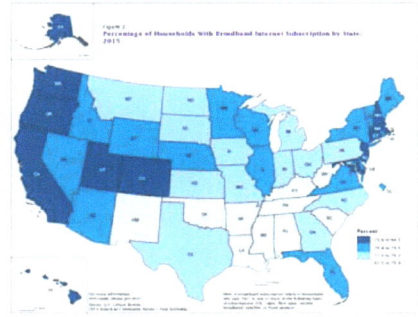

Computer and Internet Use in the United States: 2015
from US Census Bureau September, 2017
https://www.census.gov/content/dam/Census/library/publications/2017/acs/acs-37.pdf

CLICK 2 SCIENCE
click2sciencepd.org

Some groups have reached near-saturation levels for adoption of basic technologies

% of U.S. adults who say they own or use this technology

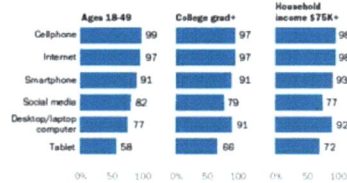

Source: Survey conducted Jan. 3-10, 2018

PEW RESEARCH CENTER

Internet, social media use and device ownership in U.S. have plateaued after years of growth, by Paul Hitlin
https://www.pewresearch.org/fact-tank/2018/09/28/internet-social-media-use-and-device-ownership-in-u-s-have-plateaued-after-years-of-growth/

CLICK 2 SCIENCE
click2sciencepd.org

The share of Americans using various technologies has stayed relatively flat since 2016

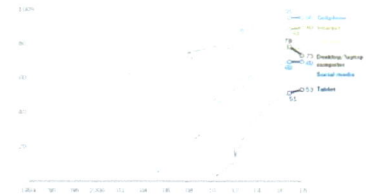

PEW RESEARCH CENTER

Internet, social media use and device ownership in U.S. have plateaued after years of growth, by Paul Hitlin
https://www.pewresearch.org/fact-tank/2018/09/28/internet-social-media-use-and-device-ownership-in-u-s-have-plateaued-after-years-of-growth/

CLICK 2 SCIENCE
click2sciencepd.org

Discuss the access your audience has to digital technology.

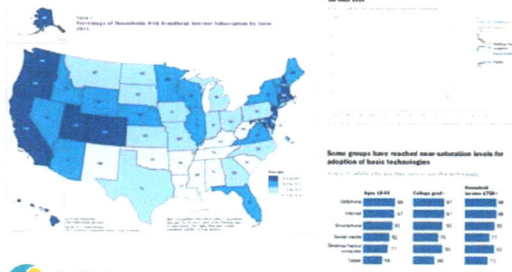

CLICK 2 SCIENCE
click2sciencepd.org

CLICK 2 SCIENCE
click2sciencepd.org

---

5 Steps to Prepare

1. Who is your audience?
2. Why is this webinar needed?
3. What type of webinar will you present?
4. How will you foster engagement?
5. Who will be on your team?

CLICK 2 SCIENCE
click2sciencepd.org

Types of Webinars

	Lecture	Interview style	Panel of Experts	Interactive
Focus on presenting information	X			X
Focus on discussion		X	X	
Focus on facilitator led activities				X
Hands-on activities or demonstrations	X			X
Diverse voices/ perspectives		X	X	
Relies on facilitator's skill	X	X	X	X
Presenters	1 presenter	Interviewer, 1 to 4 presenter	Moderator, 2 to 4 presenters	1 to 3 presenters

CLICK 2 SCIENCE
click2sciencepd.org

Fostering Engagement

- Communicate before you begin
- Prepare participants to engage
- Sandwich your content
- Use audio and visual together
- Engage expertise
- Be authentic

CLICK 2 SCIENCE

click2sciencepd.org

Gathering Your Team

1. Marketer
2. Organizer
3. Facilitator
4. Presenter(s)
5. Technicians

CLICK 2 SCIENCE

click2sciencepd.org

Preparing the Webinar Experience

CLICK 2 SCIENCE

click2sciencepd.org

Steps to Prepare

1. Who is your audience?
2. Why is this webinar needed?
3. What type of webinar will it be?
4. How will you foster engagement?
5. Who will be on your team?
6. Prepare the webinar experience.

CLICK 2 SCIENCE

click2sciencepd.org

Discussion

- Questions?
- What did I miss?
- Suggestions?

sfrerichs3@unl.edu

CLICK 2 SCIENCE

click2sciencepd.org

Level Up Your Canvas Designs: HTML and Content-Management Hacks

Steven Cain (UNL)
Tom Gibbons (UNL)
Michael Jolley (UNL)

While Learning Management Systems are great, sometimes we hit roadblocks when trying to create new or innovative design within the standard content-creation frameworks. Come learn about strategies to create interactive self-check questions in Canvas Pages using HTML, how to update content across multiple courses at once by using Canvas as a content host and other strategies to get results and information from Canvas without needing to resort to expensive external tools.

This presentation featured:

- how to deploy and update content across multiple courses without needing to host outside of Canvas.
- self-check questions in Canvas pages using HTML.
- free tools and strategies to create engaging content within the Canvas framework.

Rather than using slides we created a module in a public Canvas course that can be used as a resource. This is located at go.unl.edu/levelup. No slides were used

go.unl.edu/levelup

Overview

Introduction

Canvas can be liberating, but it can also present roadblocks. Many content creation tasks are ridiculously simple in Canvas, when compared to other LMSs. The system is also relatively easy to connect to external tools and platforms. But sometimes you need additional functionality, and the standard UI just doesn't seem to support it.

Here are three problems that we've run into as instructional designers that we struggled with in Canvas:

1. Managing common content across multiple courses, and updating it in real time.

2. Quickly gathering simple feedback from users, at the point where they are interacting with content.

3. Helping students to reinforce their learning by asking simple self-check questions without hauling out the Quiz tool.

In the next few pages, we'll share our solutions with you. We hope you find them as useful as we have.

Contact Us

- Michael Jolley mrjolley@unl.edu
- Steven Cain scain@unl.edu
- Tom Gibbons tgibbons@unl.edu

HTML Embedding

Introduction

How often have you had to recreate or replicate information in a Canvas course only to have to modify that content later and go through the entire process again?

As instructional designers at UNL, frequently we create content that gets deployed to numerous courses. When these data require updating, modifying the existing content and then remediating that content can be a time-consuming endeavor.

Although the Canvas Commons provides one possible remedy for this situation, use of the Commons requires the active participation of faculty members to update the Commons content. In addition, a lack of access to dedicated server space necessitated a more cretive solution.

Instead, we sought an answer that would enable our team to make updates and corrections to this content without the need for faculty intervention or expensive hardware. The resultant outcome takes advantage of CSS and HTML to create an elegant solution.

The solution consists of two sets of data:

1. A separate CSS file for formatting and HTML file(s) for the desired page(s)

2. Related Canvas page(s) consisting of an iFrame linking to the HTML file

Creating your HTML & CSS Files

CSS File Creation

If you plan on deploying multiple HTML pages, you may wish to use a CSS stylesheet. Using a stylesheet can ensure that any pages you create follow a consistent format. To create your CSS stylesheet, use a Text Editor program such as Text Wrangler for MacOS or Codewriter for Windows. You can also use any plain text editor as well. We created a CSS stylesheet for our purposes, designed to match the styles in Canvas, giving the appearance of being content natively built using Canvas's rich content editor, allowing it to fit seamlessly alongside other content. Here's that example CSS file:

LevelUpStylesheet.css

```
/*
Comment
*/
@import url(https://fonts.googleapis.com/css?family=Lato);
a { /* Comment */ color: /* Comment */ Red; }
body {
font-family: 'Lato', Helvetica, sans-serif;
}
h2, h3, h4, p {
font-weight: normal;
text-align: left;
text-rendering: optimizelegibility;
word-wrap: break word;
}
h2 {
font-size: 1.8em;
line-height: 1.5;
margin: 6px 0;
color: #d00000;
background-color: white;
border-top: 1px solid black;
border-bottom: 1px solid black;
padding-left: 10px;
}
```

HTML File Creation

There's a four-step process for creating the files for this solution:

1. Create the content page using an HTML text editor or the Canvas Rich Content Editor [Sample Page example]

2. If you've used the RCE, you can copy the HTML code from the RCE and paste into an HTML coding application to add any HTML coding or formatting you desire.

3. Edit the HTML code to include the html, head, CSS link, and body tags; you can modify the html code for formatting if desired

Innovation in Pedagogy and Technology Symposium, 2019

```
<!DOCTYPE html><html><head><title>Insert Page Title Here</title><link
rel="stylesheet" type="text/css" href="Insert_CSS_Reference_Here"
/></head></body>
```

4. Save the content as an HTML file locally on your computer

Hosting HTML & CSS

After we've created our CSS stylesheet and HTML file(s), we need to store those files in a publically accessible Canvas course. We'll refer to this course as the *Source Course*.

Things to consider:

- The CSS file needs to be uploaded into Canvas first so that you can include the reference to that file in the HTML file, as referenced in step 3 above.

- Both of these files should be hosted in a "source" course that is distinct from the destination(s) so that we can centrally manage them for deployment into multiple courses. Also, when you copy from one term to the next the course copy process will look for, and update, all URLs that reference the previous term and update them, so these links would break.

Generating Your iFrame Pages

iFrames are used to display a webpage within another webpage. For our purposes, we are looking to display a pre-configured webpage from an external source into our Canvas course. By doing so, we can make changes to the original files and push those changes out to any courses that might reference those pages automatically.

In its most basic form, an iFrame tag references the source of the webpage we wish to embed into our webpage.

```
<iframe src="URL"></iframe>
```

To create the iFrame code, we need to identify the item ID and the URL for the item.

Identify the item ID for the HTML file

Go to the Files section where you uploaded the files, ex. IID Sandbox-->Files-->LevelUp_Snippets

Preview the HTML file by clicking on the link, ex. "SamplePage.html"

In the preview window, note the URL for the file. It should look like this:

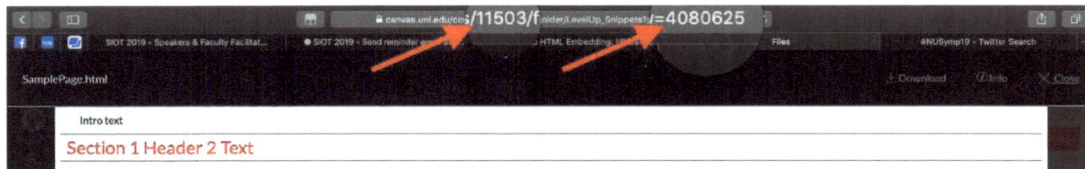

```
https://canvas.unl.edu/courses/11503/files/folder/LevelUp_Snippets?preview=4080625
```

We're interested in two numbers: 1) the course ID, and 2) the item ID. In the URL, the course ID is the first number…while the item ID is the second number.

```
https://canvas.unl.edu/courses/11503/files/folder/LevelUp_Snippets?preview=4080625
```

We need these two numbers to verify that our iFrame code is correct.

Creating the iFrame code

You may wish to create your iFrame code in a text editor first. If you recall, an iFrame simply references the URL of the webpage we wish to embed into our new page. We can also change the style of the iFrame from within the HTML code.

1. Start by writing your iframe start and end tags

   ```
   <iframe src="URL"></iframe>
   ```

2. Next, we'll add the URL of the desired file within the first iframe tag. Replace the "URL" in the iframe with the actual URL from the item preview in Canvas:

   ```
   <iframe
   src="https://canvas.unl.edu/courses/11503/files/folder/LevelUp_Snippets?preview=4080625">
   </iframe>
   ```

3. Before we proceed, we need to modify the URL to reference the file only. We want to exclude any reference to the folder in the URL, so we're going to delete the text in the URL that references the folder **"folder/LevelUp_Snippets?preview="**. Now, our iframe becomes:

   ```
   <iframe src="https://canvas.unl.edu/courses/11503/files/4080625"></iframe>
   ```

   We also need to indicate that the iframe should download the file, so we add "download" to the end of the URL:

   ```
   <iframe src="https://canvas.unl.edu/courses/11503/files/4080625/download"></iframe>
   ```

4. Now, we can add styles to our iFrame to adjust for size or other attributes. We can add the style attributes to the opening iFrame tag.

   Let's set the width of the iframe to 98% of the page width and the height to 700 px:

   ```
   <iframe style="width: 98%; min-height: 700px;"
   src="https://canvas.unl.edu/courses/11503/files/4080625/download"></iframe>
   ```

Creating the new Page in our target course

Now that we have the iframe HTML code, we can create a new page in our target Canvas course. Give the page the same title as the HTML file you uploaded into the public course, ex. "Sample Page (iFrame)"

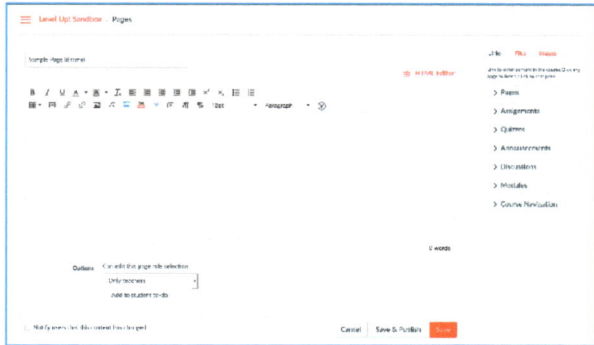

Switch from the Rich Content Editor view to the HTML Editor view

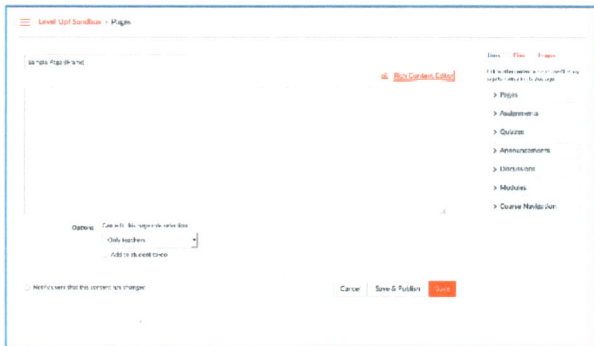

Copy the iframe code from your text editor into the HTML editor:

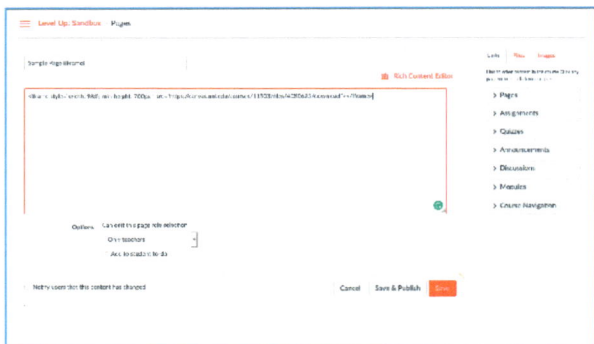

Save the page. You should now see the Sample Page from the source course embedded into the Sample Page of the target course.

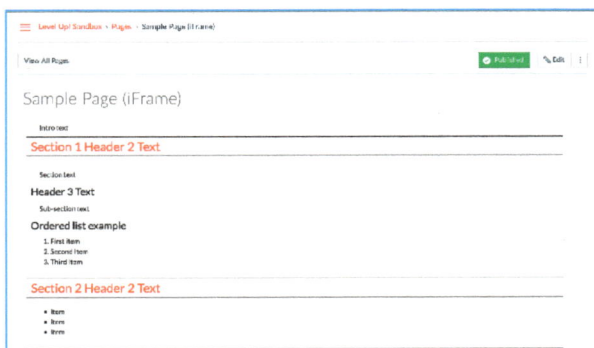

Remember, the content of this new Canvas page is the HTML page from the source course. So...if we make changes to those source HTML files, these linked pages will reflect those changes once the pages are accessed.

Things to consider: When you save the Sample Page, Canvas adds additional code to the iframe. If you edit the Sample Page in the target course and switch to the HTML editor, you should see this modified code:

```
<p><iframe style="width: 98%; min-height: 700px;" src="/courses/11503/files/4080625/download"
data-api-endpoint="https://canvas.unl.edu/api/v1/courses/11503/files/4080625" data-api-
returntype="File"></iframe></p>
```

Updating Your Source Files

When you need to make a change to your HTML or CSS document, you can upload a new version of the file into the source course. Links to previous versions of the file will point to the updated file.

How to Replace Duplicate Files in Canvas (Links to an external site.)

1. Make sure that your new file has the exact same name as your current file.

2. Upload the file into the files area in the source course.

3. When Canvas asks you if you would like to change the name of the new file or replace it, select "Replace".

Canvas will now direct all links to the previous file to the new file.

Things to consider:

- Currently, these redirects do not work unless you are logged into the Canvas instance in which the files are hosted.

Bonus: Share to the Commons

You can share your iframe pages to Canvas Commons and allow others to import your content directly into their courses. This content will then update automatically when you make changes to the source HTML files.

Things to consider:

- This will currently only work within the same instance of Canvas, due to issues with the redirects that are created when files are updated in a course.

- You can share individual pages, modules, or even complete courses. Just make sure that your HTML and CSS files are hosted in a different course than the one from which you are sharing your iframed pages. In other words, share from a target course, not a source course.

How to Share a Resource to Canvas Commons (Links to an external site.)

Google Forms

Introduction

So you're trying something new in Canvas and want to get granular page-by-page feedback. In my case this originally came about with the idea of blocks of content that could be centrally managed, but still let us receive feedback so we could know when changes needed to be made. I ended up using it in a pilot course to see which types of content and engagement resonated most with my audience. It provided a clear picture of what was working with my students and where I should be focusing my efforts for additional improvement.

Keep in mind that the processes outlined on the page are very character-dependent. Relying on manual data entry is both laborious and likely to cause issues in the process. Aside from creating your form for the first time, you shouldn't have to do much manual typing for this process.

Setting up the Form

Here's a resource to help you get started with Google Forms (Links to an external site.), if you've never used it before.

For this type of project my recommendation is to keep the form *incredibly simple*. The more questions you have to auto-fill with your link, the more complicated creating the URLs will be.

Once your form is created, use the Get Pre-filled Link tool to populate fields and generate a link.

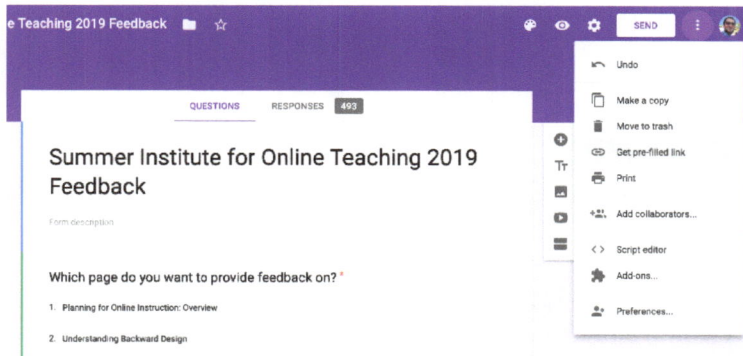

Deconstructing and Modifying the URL

I selected the first page, and that it was helpful.

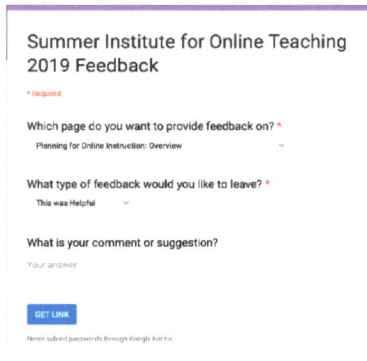

And here's the pre-filled link that I got from Google Forms:

```
https://docs.google.com/forms/d/e/1FAIpQLSc0jKyR0BJv3lzBXGeRB_Dvq1U4_yU7s3z28
Ute27rHz3WLcw/viewform?usp=pp_url&entry.1700942623=Planning+for+Online+Instru
ction:+Overview&entry.406679041=This+was+Helpful
```

If we look closely at the URL's structure there are a few sections that stand out.

```
https://docs.google.com/forms/d/e/1FAIpQLSc0jKyR0BJv3lzBXGeRB_Dvq1U4_yU7s3z28
Ute27rHz3WLcw/viewform?usp=pp_url&entry.1700942623=
```

```
Planning+for+Online+Instruction:+Overview
```

```
&entry.406679041=
```

```
This+was+Helpful
```

And now let's look at one more, for comparison

```
https://docs.google.com/forms/d/e/1FAIpQLScOjKyROBJv3lzBXGeRB_Dvq1U4_yU7s3z28
Ute27rHz3WLcw/viewform?usp=pp_url&entry.1700942623=Course+Organization+%26+St
ructure&entry.406679041=Comment/Suggestion
```

And if we deconstruct that one...

```
https://docs.google.com/forms/d/e/1FAIpQLScOjKyROBJv3lzBXGeRB_Dvq1U4_yU7s3z28
Ute27rHz3WLcw/viewform?usp=pp_url&entry.1700942623=
```

Course+Organization+%26+Structure

```
&entry.406679041=
```

Comment/Suggestion

You'll notice that four of these lines are identical, but two are different. I've bolded the different pieces. I also bolded the viewform section of each of these, which leads into...

To make the form auto-submit

To remove as many barriers as possible, I wanted the helpful and not helpful buttons to auto-submit. In the case of "This was helpful" and "This was not helpful" you may want to make the form auto-submit. This means as soon as the link is clicked the response is received and the end-user can just close the window. All you need to do is change **viewform** to **formResponse**:

```
https://docs.google.com/forms/d/e/1FAIpQLScOjKyROBJv3lzBXGeRB_Dvq1U4_yU7s3z28
Ute27rHz3WLcw/formResponse?usp=pp_url&entry.1700942623=
```

```
Planning+for+Online+Instruction:+Overview
```

```
&entry.406679041=
```

```
This+was+Helpful
```

Generating Links en Masse

We'll start with a column in a spreadsheet that includes the name of every page for which you want to generate links. To get this list from the Google form you can use the print function to get a clean list that can be copied and pasted into a spreadsheet.

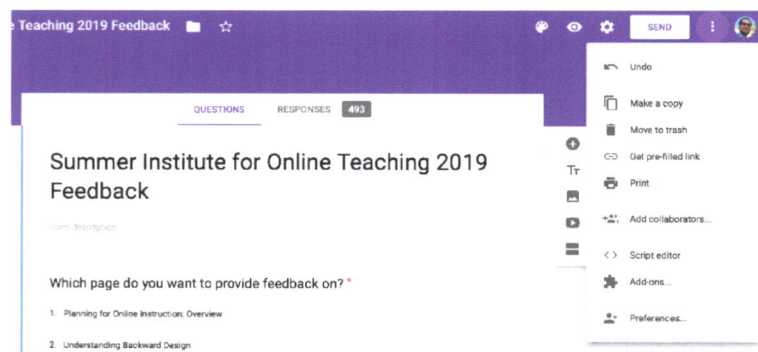

First, we need to clean up the names to make them function as part of the URL. Use **find & replace** to change all spaces to "+" and special characters to their appropriate hex codes (Links to an external site.).

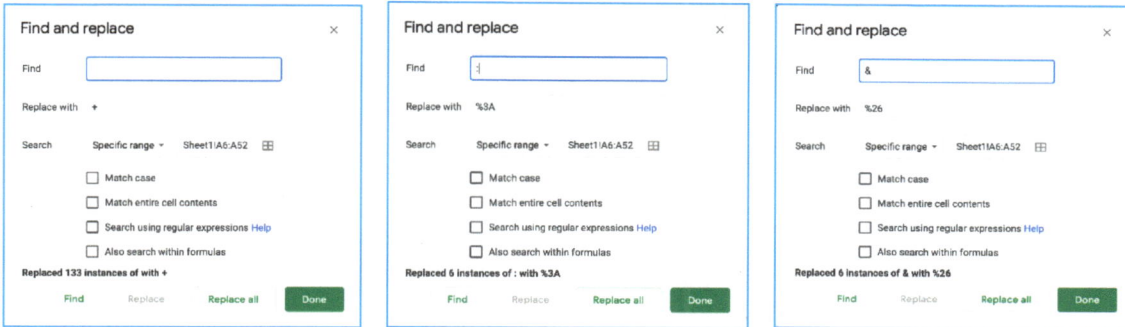

You'll notice some of the characters may not **need** to be changed (the colon in *Planning for Online Instruction: Overview* didn't, for example), but changing them won't cause the link to break, whereas not changing one that needs to be changed definitely can break the link.

Do the same for the types of responses that correspond with each page (i.e. Helpful, Not Helpful, Comment/Suggestion).

Now we can use the CONCATENATE function to merge these pieces with the URL stem pieces.

```
=CONCATENATE(OpeningURLString, ModifiedPageName, MiddleURLString, ModifiedFeedbackType)
```

To copy this into every page we just need to set a fixed reference for everything other than the ModifiedPageName. To set a fixed reference we just add $s in front of the cell references that we want to keep locked in place.

So *=CONCATENATE(A1,A6,A3,B5)* becomes *=CONCATENATE(A1,A6,A3,B5)*. Do this for the first value of each response type (don't forget to use the *viewform* stem instead of the *formResponse* stem if you don't want it to auto-submit), then just copy them down the entire column!

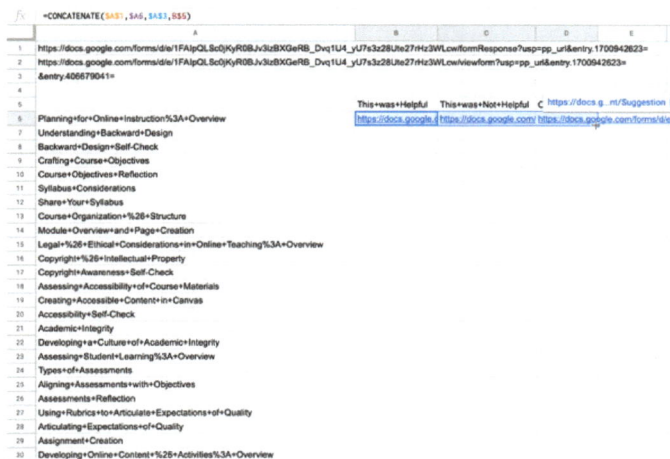

Creating HTML Snippets

Here's the snippet I used. If you wanted to give it a different style, feel free to play around, but you will likely have the same general structure. You can see highlighted in the code the different areas where I the URLs we generated in the last step are inserted.

```
<hr /><h2 style="text-align: center;"><span>Feedback</span></h2><p
style="margin-left: auto; margin-right: auto; text-align: center;"><a
style="width: 29%; display: inline-block; border-color: #c3e6cb; background-
color: #c3e6cb; color: #155724; text-align: center; padding: 5px;" href="
```

_Insert_your_helpful_link_here_

```
" target="_blank" rel="noopener">This Was Helpful</a><a style="display:
inline-block; border-color: #f8d7da; background-color: #f8d7da; color:
#990000; text-align: center; padding: 5px; margin-left: 1%; margin-right: 1%;
width: 29%;" href="
```

_Insert_your_not_helpful_link_here_

```
" target="_blank" rel="noopener">This Wasn't Helpful</a><a style="display:
inline-block; border-color: #ffeeba; background-color: #ffeeba; color:
#856404; text-align: center; padding: 5px; width: 29%;" href="
```

_Insert_your_Comment/Suggestion_link_here_

```
" target="_blank" rel="noopener">Suggest an Update</a></p>
```

Feedback

This Was Helpful (Links to an external site.)This Wasn't Helpful (Links to an external site.) Suggest an Update (Links to an external site.)

In fact, you could continue with the CONCATENATE function to generate unique HTML snippets for each page! Just be careful to include all the standardized characters that you need (quotations are required around URLs) and don't forget to lock in place the standardized pieces of the HTML snippet. I've broken up the snippet from above into the fixed and differentiating pieces to help you out.

```
<hr /><h2 style="text-align: center;"><span>Feedback</span></h2><p
style="margin-left: auto; margin-right: auto; text-align: center;"><a
style="width: 29%; display: inline-block; border-color: #c3e6cb; background-
color: #c3e6cb; color: #155724; text-align: center; padding: 5px;" href="
```

_Insert_your_helpful_link_here_

```
" target="_blank" rel="noopener"><i class="fas fa-thumbs-up" aria-
hidden="true"> </i> This Was Helpful</a><a style="display: inline-block;
```

```
border-color: #f8d7da; background-color: #f8d7da; color: #990000; text-align:
center; padding: 5px; margin-left: 1%; margin-right: 1%; width: 29%;" href="
```

Insert_your_not_helpful_link_here

```
" target="_blank" rel="noopener"><i class="fas fa-thumbs-down" aria-
hidden="true"> </i> This Wasn't Helpful</a><a style="display: inline-block;
border-color: #ffeeba; background-color: #ffeeba; color: #856404; text-align:
center; padding: 5px; width: 29%;" href="
```

Insert_your_Comment/Suggestion_link_here

```
" target="_blank" rel="noopener"><i class="fas fa-file-signature" aria-
hidden="true"> </i> Suggest an Update</a></p>
```

Now you can copy/paste these snippets into Canvas pages, or even more conveniently, use the Embed Media function within Canvas to drop it into place.

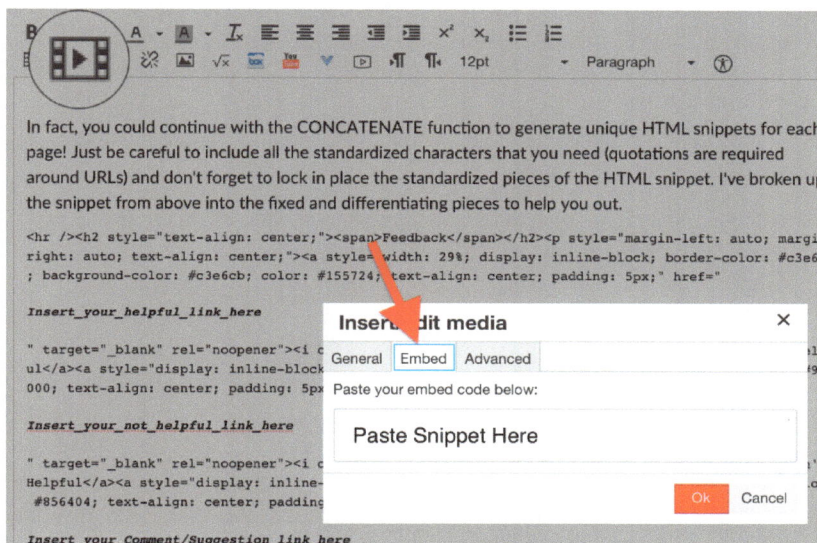

Word of Warning

As valuable as the Canvas link validator is, please note that when it runs it pings every URL in the course. If you use this method and then run the link validator it will report that each page was both helpful and not helpful. **Be sure to delete all responses after running the link validator and before the start of the course itself.**

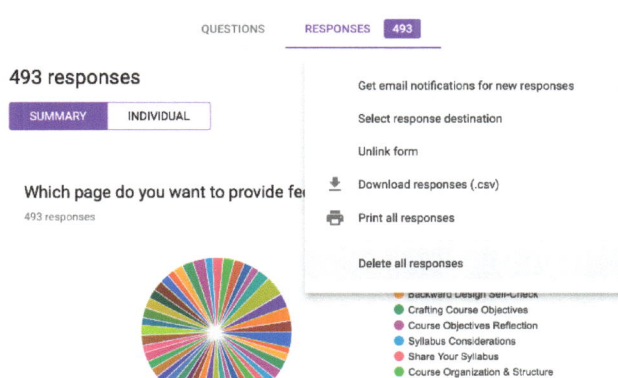

Tying it all together

Using the HTML embedding strategy in conjunction with leveled up Google Forms also allows us to do some interesting things, such as creating this educational reminder snippet that could be placed in any number of classes, centrally managed, and with feedback buttons that clearly refer to the current version.

Interactive Self-Check

Introduction

When creating lessons in Canvas, many content-creators want to include practice questions, so that students can check their knowledge and comprehension along the way. The Quiz tool in Canvas provides one method of doing this, but has three significant limitations:

1. It lacks granularity--you can't ask questions that are distributed within the flow of a content page. You have to create an additional content object in the course.

2. It disrupts the flow of the content--the quiz exists outside of the context of the content page. Inserting a quiz after every content page would quickly become cumbersome.

3. It's difficult to use in publicly visible resource courses, for which the participants are not enrolled in the course. When the self-enroll function has been disabled in Canvas, creating automated feedback mechanisms for self-paced or JiT training materials becomes a challenge.

It's this third condition that led to the investigation of alternate solutions. I was working with ITS to construct a resource for accessibility in Canvas that would be open and available to anyone. I ultimately settled on this self-check solution.

Example and Complete Code File

Answer the following questions:

Q1

When working with pigments, what color do you get when you mix red with blue?

Q2

Solve for x using the quadratic equation:

$$10 x 2 = 6 + 9 x$$

Download the code for everything inside the box.

togglebuttoncode.txt

Code Walkthrough

Creating messages that are only visible in the app

This will hide all content contained between the div tags from browser-based viewports.

Because the mobile apps don't load the styles necessary to make this work, it's necessary to include a message that's only visible in the apps indicating that the interactive content doesn't function. These buttons do work in mobile **browsers**, just not the apps. So, we want to hide this message for desktop, tablet and phone **browsers**.

```
<div class="hidden-desktop hidden-tablet hidden-phone" style="background-color: #d00000;">
<p style="color: white; padding: 1em;">The following content will not function in the mobile app. To access the interactive content, please login to Canvas using a web browser.</p>
</div>
```

Here's what this looks like in the app:

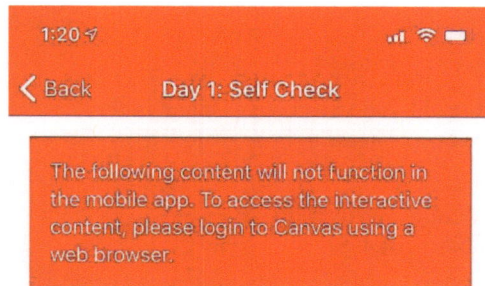

Answer the following questions:

Q1

Why is using Headers better than just increasing the text size?

Reveal the answer

Q2

Why should you use descriptive names in links?

Question Text

The question text is styled as a regular paragraph, or in any other way you would like. You may even want to come up with a specially-styled div that you use throughout your course to wrap all of your self-check questions to highlight them within your content.

```
<p>When working with pigments, what color do you get when you mix red with blue?</p>
```

Button

This next block of code creates the button that will be used to reveal the hidden answer text.

```
<p><span class="element_toggler btn btn-primary" role="button" aria-controls="group_2" aria-label="Toggler toggle list visibility" aria-expanded="false">Reveal the answer</span></p>
```

The first part of the span tag calls the toggler button from the Canvas CSS and designates it's role on the page.

```
<span class="element_toggler btn btn-primary" role="button"
```

The next part provides an aria controls designation for the button. This is one half of a pair of attributes that associates the button with its respective answer content. The name inside the quotation marks must be the same as the name used in the div id for the answer text.

```
aria-controls="group_2"
```

The next section of the tag provides the aria label--a name for the function of the button, its purpose--and then the state of the button on page load.

```
aria-label="Toggler toggle list visibility" aria-expanded="false">
```

Finally, the text of the button is placed, along with the close span tag. The paragraph tags are additions from the Canvas Rich Content Editor.

```
Reveal the answer</span></p>
```

Answer Text

The next important bit of this is the styling for the answer text.

```
<div id="group_2" class="content-box" style="display: none;">Purple</div>
```

The first part opens the div, then names the id for the answer. This id must match the aria-controls name in the button code, and this must be a unique pairing for the page if you want the content items to be revealed independently.

If you wanted a single button to control multiple reveals, you would give all of the answer divs the same id as the aria-controls name for the button. This raises some questions about accessibility, though.

```
<div id="group_2"
```

The next part loads the class for this text from the Canvas CSS.

```
class="content-box"
```

Then the code sets the visibility of the answer text to hidden on page load.

```
style="display: none;">
```

Finally, the answer text and the closed div tag.

```
Purple</div>
```

Possible Use Cases

- Just in Time, open, or self-paced training

- Longer pages of content

- Critical content that needs to be reinforced

- Possible question types:
  - Identification
    - Visual/multimedia
    - language acquisition
  - Definitions
  - Multiple choice
  - True/False
  - Jokes with punchlines

Why did the chicken cross the road?

Where this Code Breaks Down

The Mobile App

As mentioned, above, the mobile app doesn't load the styles necessary to make this strategy work. You'll need to advise users to use a web browser to access Canvas to use this strategy effectively.

If you're building content that you know will be primarily accessed using small form-factor mobile devices, this may not be the right approach. The mobile browser experience in Canvas can be a little fiddly on smaller viewports.

Browser-Based Tab Navigation

As of now, I can't figure out a way to tab to and actuate the button using a keyboard. It's possible--even likely--that screen-reader users may be fine with the current strategy, given the aria labels. But I haven't tested it, and I'm reaching the fringes of what I know about behind-the-code accessibility.

It's important to be aware that keyboard- or voice-navigation users may be excluded from content deployed in this way. It's something that should be tested before wide-scale deployment.

Invisible Editing

For the sections that are set to be hidden in the desktop, or hidden with the toggler, you'll need to edit the content in the HTML Editor.

OR, if you need some of the Rich Content Editor tools, like the equation editor, you can approach it one of two ways:

- Create your content in the RCE in a visible place, then switch to the HTML Editor and then copy and paste your code into the hidden div.

- Or, change the code that hides the content while you edit, then replace it when you're done.

  - For the warning message you may need to remove **hidden-desktop** from the div class. Lately, it's been visible in the edit window, but it wasn't when I started playing with this strategy.

  - For the togglers, you'll need to make two changes

    - change **aria-expanded="*false*"** to **aria-expanded="*true*"** in the button tag

    - delete **style="display: none;"** from the div that contains the answer content for the toggler

Transitioning to the Hybrid Model: Preparation to Ensure High-Quality Distance Education

Melissa Cast-Brede, Ph.D. (UNO)

Sarah K. Edwards, Ph.D. (UNO)

Erica Rose (UNO)

The Teacher Education Department at the University of Nebraska at Omaha implemented a two-year process of thoughtful discussion for intentional decision-making focused on faculty voice and ownership. This intentional approach allowed all faculty to develop hybrid graduate programs that maintained quality instruction and prioritized students. Through the identification of early adopters and the effective use of departmental and campus resources, a series of professional development opportunities were developed to assist faculty in course design and introduce instructional tools that aid content distribution and student engagement. Follow up initiatives surrounding the integration of educational technology and additional student engagement professional development worked to ensure sustainable practices in the delivery of quality instruction. This presentation will review the decision processes and benchmarks, professional development approach, and campus partner supports that led to a successful transition, as well as discuss the next steps in maintaining the momentum.

This presentation featured:

- processes for fostering ownership in blended learning.
- tips on how to manage resources for effective professional development.
- strategies for maintaining the momentum for growing distance programs.

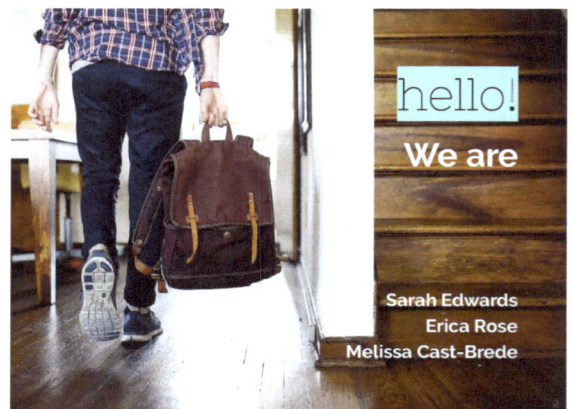

Slide 1

1. Decisions and Benchmarks

Slide 2

Impact of Hybrid Model

	2016	2019
Number of Courses in Graduate Programs	13 face to face 17 hybrid	5 face to face 23 hybrid 2 online
Enrollment Trends in Graduate Programs	245	291
Faculty Evaluation Numbers	4.1 course avg 4.24 instructor avg	4.27 course avg 4.36 instructor avg

Slide 3

Build from Success

Highlight Program

↓

Leaders and Early Adopters

↓

TED Talks

Slide 4

2. Partners and Support Systems

Slide 5

Find Them!

Ask Them!

Slide 6

UNO Center for Faculty Excellence

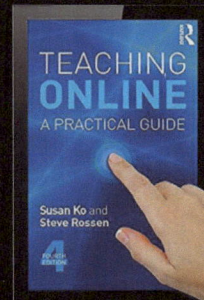

TEACHING ONLINE
A PRACTICAL GUIDE

Susan Ko and Steve Rossen

Ko, S., & Rossen, S. (2017). *Teaching online: A practical guide* (Fourth ed.). New York: Routledge.

Slide 7

Course Planning Template

Unit / Week	Objectives	Targeted Skills	Course Materials (Instructor-Generated Content, Communications, Feedback)	Class Interaction & Activities	Readings & Resources	Assessments (Assignments, Exams, Projects, etc.)

Ko, S., & Rossen, S. (2017). *Teaching online: A practical guide* (Fourth ed.). New York: Routledge.

Slide 8

Course *Redesign* Template

Learning Objective	New Learning Objective	Previous Course Materials	New Course Materials	Previous Class Interaction & Activities	New Class Interaction & Activities	Previous Assessment	New Assessment

Ko, S., & Rossen, S. (2017). *Teaching online: A practical guide* (Fourth ed.). New York: Routledge.

Instructional Designers
- Tech tools
- Panopto (now VidGrid)
- Zoom
- Brainstorming activities

UNO Office of Digital Learning

UNO Criss Library ~ Education Librarian

LibGuides
Streaming Videos
Article Persistent Links
Electronic Reserve
Video Tutorials

COE Tech Office

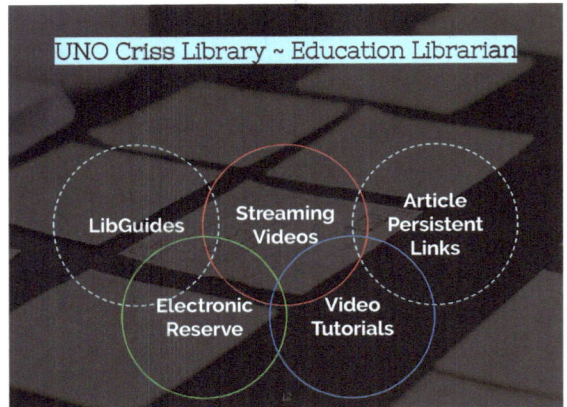

Coordinating Equipment Needs
Virtual Guest Speakers

3. Professional Development Approach

TED Talks Student Engagement 3.0
Fall, 2016 – 2019

"Engagement results when students involvement in learning contributes to their learning and sustains their involvement in the course" (Meyer, 2007, p.6).

Framework

- **Teaching Presence**
  - Design & Organization; Facilitation; Direct Instruction
- **Social Presence**
  - Affective expression; Open communication; Group cohesion
- **Cognitive Presence**
  - Triggering event; Exploration; Integration; Resolution
- **Learning Presence** (proposed addition to CoI)
  - Effective Learner Behavior

Community of Inquiry

Social Presence
Supporting Discourse
Cognitive Presence
Educational Experience
Setting Climate
Selecting Content
Teaching Presence
Communication Medium

Polling the crowd; finding our experts

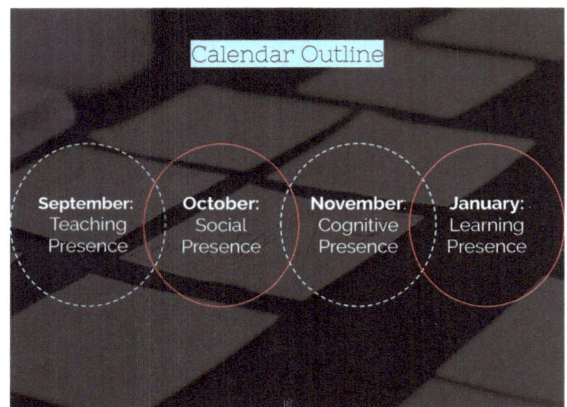

Calendar Outline

September: Teaching Presence
October: Social Presence
November: Cognitive Presence
January: Learning Presence

Format

Framework

Examine the framework

Define the "presence"

Examples

Expert Presentations

3-4 presenters

Discussion of strategy

Demonstration/examples

Universal application

Small group discussion

Share engagement strategies

Discuss something to try in the future

Gather and share resources

Canvas Page (TED Resources)

---

4. Maintaining Momentum

Canvas Stipends

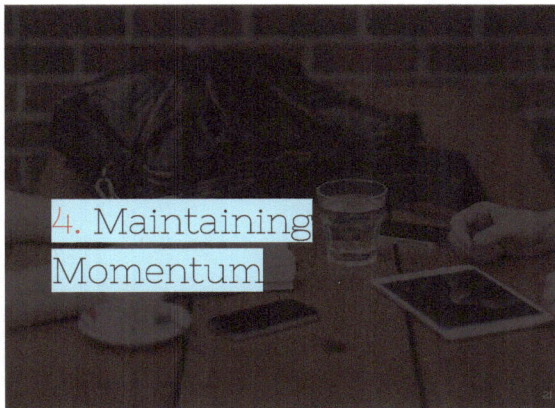

1. Posted Office Hours
2. Syllabus- official, current syllabus is posted for students
3. Course Schedule- posted for students
4. Gradebook- in use for students
5. Closed captioned VidGrid video
6. LiveText link for a specific assignment
7. View course via the student view: Reflection

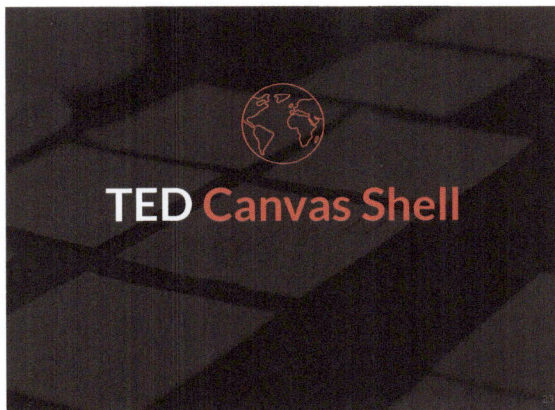

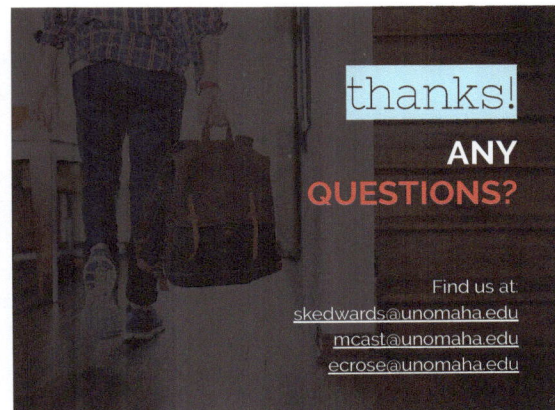

TED **Canvas Shell**

Before we began...
Polling and voting

OER: Open Educational Resources

- Fall Pilot
  - 5-6 faculty
  - OER Coordinator
  - Education Librarian
- Bring it back

---

thanks!

ANY QUESTIONS?

Find us at:
skedwards@unomaha.edu
mcast@unomaha.edu
ecrose@unomaha.edu

Ask the Pros: An Interactive Discussion with a Futurist & a Humanist

Bryan Alexander, Ph.D.

Tanya Joosten, Ph.D.

Two of online and higher education's leading thinkers discussed their key takeaways from the day as well as their thoughts on up-to-the-minute trends and breaking news. Attendees had the opportunity to ask questions by using the Slido app or #NUSymp19 hashtag on Twitter.

Closing Remarks

Mark Askren,
Vice Chancellor for IT and CIO, University of Nebraska

Committees

Executive Committee

Mark Askren
Vice President for Information Technology
and Chief Information Officer
mark.askren@nebraska.edu

Mary Niemiec
Associate Vice President Distance Education,
Director University of Nebraska Online
mniemiec@nebraska.edu

Programming Committee

Leona Barratt
Academic Technology Training Associate, UNCA
leona.barratt@nebraska.edu

Rich Murch-Schafer
Lead Instructional Technologist, UNO
rmurch-shafer@nebraska.edu

Tanya Custer
Assistant Professor, Distance Education Coordinator, UNMC
tcuster@unmc.edu

Eyde Olson
Instructional Design Technology Specialist, UNL
eolson2@unl.edu

Melissa Diers
Senior Instructional Designer, UNMC
mdiers@unmc.edu

Deb Schroeder
Assistant Vice President & CIO, ITS
schroederd@unk.edu

Elizabeth Leader Janssen
Associate Professor, UNO
eleaderjanssen@unomaha.edu

Eric Tenkorang
Instructional Designer, UNK
Tenkorange2@unk.edu

Diane Kortus
Learning & Development Specialist, UNK
diane.kortus@unmc.edu

Yaoling Wang
Instructional Designer, UNK
Yaoling.wang@unl.edu

Steve McGahan
Associate Director & Instructional Design Specialist, UNK
mcgahansj@unk.edu

Eric Tenkorang
Instructional Designer, UNK
Tenkorange2@unk.edu

Planning Committee

Todd Karr
Assistant Director, University of Nebraska Online
tkarr@nebraska.edu

Kenna Schneringer
Marketing Coordinator, University of Nebraska Online
kschneringer@nebraska.edu

Laura Wiese
Director of Marketing & Communication, University of
Nebraska Online
ljwiese@nebraska.edu

Cortney Kirby
Marketing Coordinator, University of Nebraska Online
ckirby@nebraska.edu

Jill Bertsch
Associate Director of Marketing & Communication,
University of Nebraska Online
ljwiese@nebraska.edu

Elizabeth Thorne
Marketing Coordinator, University of Nebraska Online
ethorne@nebraska.edu

Colleen Huls
Human Resources Coordinator, NU ITS – UNL
chuls1@unl.edu

www.ingramcontent.com/pod-product-compliance
Lightning Source LLC
Chambersburg PA
CBHW041717210326
41598CB00007B/687